film: Mindwalk

SCIENCE

ON STAGE

SCIENCE

ON STAGE

From *Doctor Faustus*
to *Copenhagen*

KIRSTEN SHEPHERD-BARR

PRINCETON UNIVERSITY PRESS
PRINCETON AND OXFORD

Library of Congress Cataloging-in-Publication Data

Shepherd-Barr, Kirsten, 1966–
Science on stage : from Doctor Faustus to Copenhagen / Kirsten Shepherd-Barr.
p.   cm.
Includes bibliographical references and index.
ISBN-13: 978-0-691-12150-5 (hardcover : alk. paper)
ISBN-10: 0-691-12150-8 (hardcover : alk. paper)
1. Drama—20th century—History and criticism. 2. Science in literature.
3. Scientists in literature. I. Title.
PN1650.S34S54   2006
809.2′9360904—dc22   2005027627

British Library Cataloging-in-Publication Data is available

This book has been composed in Goudy

Printed on acid-free paper. ∞

pup.princeton.edu

Printed in the United States of America

1   3   5   7   9   10   8   6   4   2

Contents

············

Acknowledgments

· ·

This book could not have been written without the help and support of so many people and institutions. I wish to thank my colleagues at North Carolina State University for their support and encouragement, in particular Mary Helen Thuente and the Dean of the College of Humanities and Social Sciences for awarding me a summer stipend, as well as extended departmental leave, which helped me to complete the book. I am also grateful to the Honors Program at NC State for encouraging me to implement new interdisciplinary courses on science and theater, out of which this book has grown.

During the writing of this book, Clare Hall, Cambridge, became a home away from home and provided invaluable resources for which I am extremely grateful. In Cambridge I also attended a colloquium on *Copenhagen* in Jesus College in November 2003, for which I would like to thank its organizer, John Cornwell. Thanks to professor Marcial and Maria-Louisa Echenique of Farm Hall for graciously allowing me to visit, and to Kate Price, Liliane Campos, and other members of the Science and Literature Reading Group at Darwin College, Cambridge. Thanks to Dame Professor Gillian Beer for her ongoing encouragement and her example. My gratitude also goes to my new colleagues in the Department of Drama and Theatre Arts at the University of Birmingham, for helping me adjust to the new job and its challenges while also trying to finish the book. I wish to thank Pino Donghi of the Sigma Tau Foundation for enabling me to see *Infinities* and interview the director. I wish to thank my students, on whom much of this material has been tested, both at NC State and at Birmingham. They have been guinea pigs in the process of developing this work, and I have learned a tremendous amount from them.

The manuscript has benefited from having astute readers from both science and the humanities at several stages of its completion. Among the scientists who have given invaluable insights and advice, I would like especially to thank Harry Lustig, Eugen Merzbacher, John Barrow, Alain Prochiantz, Richard Hyder, Jeremy Bernstein, Gordon M. Shepherd, and Alastair J. Barr. I have also benefited from stimulating exchanges with Carl Djerassi. Other enormously helpful readers of the book in its early stages include in particular Michael Vanden Heuvel, whose insights and suggestions have been indispensable. Any errors of fact are entirely my own. I owe a special debt of gratitude to several people who have kindly provided translations of passages from German and French into English: Helen Hyder, Grethe Shepherd, and Lisbeth Shepherd.

Some of the material in the book has appeared in earlier versions in other publications, and I would like to thank the following journals for permission to use portions of that work: *Physics World*, *SIAM News*, *Gramma*, and *American Scientist*. I am also grateful to Michael Frayn for permission to quote him.

I have been fortunate to have a wonderful editor, Vickie Kearn, to whom I extend my warmest thanks, and to her assistants Alycia Somers and Susan Berezin. Finally, I would like to thank my parents for their ongoing encouragement and support, my husband, Alastair, and my children, Graham and Callum. This book is for them.

Introduction

• • • • • • • • • • • • • • • • •

Michael Frayn's 1998 play *Copenhagen*, about a secret meeting between the physicists Niels Bohr and Werner Heisenberg during World War II, brought science to the theater in an unprecedented way. The play proved hugely popular and led a wave of "science playwriting" that moved one critic to proclaim that "science is becoming the hottest topic in theater today, so much so that it's identifiable as a millennial phenomenon on the English-speaking stage."[1] Yet *Copenhagen* is just one of the latest in a long line of plays that take science as their subject matter and scientists as their protagonists. For centuries (see appendix, "Four Centuries of Science Plays"), science and theater have enjoyed a fruitful intersection in the form of dramas that utilize scientific ideas or feature scientists at their center, from Marlowe's *Doctor Faustus* and Jonson's *The Alchemist* to Shaw's *The Doctor's Dilemma* and Brecht's *Life of Galileo*.

But over the past decade, a surge of new plays about science has created a true phenomenon. Plays like *Copenhagen, Arcadia, Wit, After Darwin, A Number, An Experiment with an Air-Pump,* and *Molly Sweeney* have made the stage a major forum for the exploration of scientific ideas and given the theater pride of place as the site of substantive interaction between the hard sciences and the humanities. No other genre or art form has seen such a powerful merging of the two cultures of science and humanities. Yet to date this phenomenon has received relatively little sustained critical attention, which is surprising at a time when, as the critic Susan Stanford Friedman writes, "the borders between disciplines and divisions of knowledge have become increasingly porous . . . with the re-legitimation of interdisciplinarity."[2] *Science on Stage* seeks to provide the first full-length analysis of this interdisciplinary phenomenon we are conveniently, if not completely satisfactorily, calling science plays.[3]

Science plays span many cultures as well as periods. Predominantly British, they also come from America, France, Germany, Switzerland, Greece, and other mainly European countries. They encompass a remarkable range of scientific fields and ideas; neurology, physics, biology, evolution, genetics, reproductive technology, epidemiology, biochemistry, mathematics, and astronomy are just a few. Since a comprehensive survey of all science plays is impossible, I will look at a few of the most significant, representative examples to get a sense of how this long-standing, cross-cultural, and interdisciplinary genre has evolved. For my purposes, they will be shown to share certain critical features: a casting of the scientist as hero or villain (or sometimes both), a direct engagement with "real" scientific ideas, a complex ethical discussion, and an interdependence of form and content that often relies on performance to convey the science.

Also worth noting about these plays is their unusually wide formal spectrum; the sheer range of styles and structures of contemporary science plays is dazzling. They span straight realism, documentary theater, and Brechtian theater, and they also generate new forms like the spare, pared-down *Copenhagen*, with just a few actors and chairs on an otherwise bare stage and a setting that is halfway between realism and the supernatural. A good many contemporary science plays utilize alternating time periods, usually with actors doubling roles in both. There seems to be an impulse on the part of the science playwright to call on the audience's imagination more than is usually done in the theater. Perhaps this formal innovation and experimentation relates to the science playwright's special problem of exposition: it is not just about getting a story across, but presenting a set of ideas that can be quite complicated to explain.

One of the striking things about many contemporary science plays in particular is their combination of textual richness and scenic restraint. There is a marked tendency to de-emphasize spectacle: most of them feature a few actors on a bare or minimally decorated stage and a heavy amount of dialogue. This foregrounds both the text and the actors' bodies. Examples of this are Caryl Churchill's *A Number* and Michael Frayn's *Copenhagen*, perhaps unprecedented in the amount and degree of difficulty of the line memorization for the actors, and staged with only two or three actors and a matching number of chairs. Brian Friel's *Molly Sweeney* also features three actors and three chairs. Timberlake Wertenbaker's *After Darwin*, set in two time periods with four actors, some scenery, and minimal props, looks almost extravagant next to these. Yet "boring" has been far from the response of critics and audiences to these visually spare works; in fact, one of the aims of this book is to try to understand why this combination of visual minimalism and textual abundance has proved so successful in an art form that more often sees the inverse combination as its norm for success.

The opening chapter on the tradition of science plays contextualizes the recent explosion of plays about science. The book recognizes this dramatic genre, and does so from the perspective of drama and theater, with an emphasis on performance. Drawing on interviews with playwrights and directors and on recent work in theater studies, especially performativity and the newest thinking about speech-act theory, *Science on Stage* illuminates the forces that have combined to produce this phenomenon of modern science plays, as well as the receptive audiences for them.

There is something very appealing in the quest of the scientist; audiences are drawn to the lone crusader's neoromantic pursuit of truth, knowledge, and beauty. Yet, as scientists become increasingly dependent on the public imagination for reliable truths, and as those truths themselves become troubling in their potential to challenge our fundamental conceptions of humanity, scientific power starts to become problematic; plays like Howard Brenton's *The Genius*, Shelagh Stephenson's *An Experiment with an Air-Pump*, and Caryl Churchill's *A Number* raise questions about the public responsibility of the scientist and the nature of his or her pursuits. This is one of the paradoxes that *Science on Stage* will explore in the attempt to understand the remarkable new wave of science plays over the last decade. Chapter 2 explores the appeal of science plays by addressing such questions as the following:

What accounts for the number and quality of plays about science and scientists in the past one hundred years? What accounts for this phenomenon, especially the spate of recent science plays?

Are developments in science a topic in drama because of our tendency to "dramatize" processes of discovery, to cast investigators as heroes?

Conversely, do plays contribute to the telescoping of those processes of discovery into dramatic moments of insight?

Do some plays try to debunk this mythology of scientific heroism to which other, more reactionary dramas lend support?

How has theater been able to overcome its inherent constraints—limits on space and effects, emphasis on action—to become a successful means of inquiry into the nature and repercussions of scientific discovery?

Are audiences attracted to science plays despite or because of their often difficult subject matter? Is the answer merely a socioeconomic one relating to level of audience education and wealth, or are other factors involved?

What kinds of debates and controversies have arisen over the use of science on the stage?

"The only thing really worth teleporting is information" - Physics of Star trek - History channel Theory of Space

How do science plays help to bridge the gap that still persists between the two cultures?

In what ways are science plays performative?

And what separates the best ones from the less successful plays?

The sheer proliferation of science plays over the last two decades has indicated a need for taxonomy. There have been attempts to categorize science plays according to distinct types—for example, "plays about social implications of science, plays about scientists as people, and plays which center on science itself."[4] Such distinctions cannot hold, however, not just because there is a great deal of overlap but because they fail to take theatricality into account; they focus solely on thematic content. I want instead to suggest some alternative types or categories that revolve around theater as much as science. Lots of "science plays" are by playwrights who are not scientists but have become interested in using science in their work and have researched a particular scientific idea or figure. Examples of this current are Frayn and Stoppard, Brecht and Friel, Wertenbaker and Churchill. Conversely, there are plays by scientists who are fascinated by the potential of the stage to convey scientific ideas. Examples of this type of play are those by Carl Djerassi and Elizabeth Burns. A third type of science play utilizes documentary theater; several examples of such science "docudramas" are analyzed in chapter 3. Finally, a new trend in science playwriting is the collaboration between scientists and theater directors; examples are Peter Brook and Marie-Hélène Estienne, Luca Ronconi and John Barrow, and Alain Prochiantz and Jean-François Peyret. This latter trend is especially interesting and innovative, and will be explored in the conclusion. The book also exposes the deep fault lines dividing these various groups of science plays and their advocates. The vigorous debates that have arisen about what constitutes a good science play and what is fair use of science and history are a sure sign that the field of theater and science not only exists but is in robust health.

Whatever the particular type or category, most science plays are, quite simply, "attempts to investigate human problems by reference to scientific ideas."[5] Many have enjoyed great stage success as well, both with audiences and with critics; several have won prestigious prizes like the Tony Award and the Pulitzer. A book about theater and science is timely, since both fields have long borrowed from one another for metaphoric explanations of what they are and do. On the one hand are the scientists using the language of the theater, from physicists likening quantum mechanics to the stage,[6] to the cognitive neuroscientist Bernard Baars in his book *In*

metamaterial for cloaking
absorbing is easier than deflecting

the Theater of Consciousness: The Workspace of the Mind (1997). On the other hand are the nonscientists who find the language of science appealing, as in David Lodge's recent essays and novels suggesting that the novel is the perfect expression of consciousness and that novelists therefore hold the key to defining and locating consciousness; and William Demastes's Staging Consciousness: Theater and the Materialization of Mind (2002), which links modern drama with the science of consciousness. Science has also become increasingly apt as a metaphor for theatrical performance itself. "Putting on a play itself is a sort of scientific experiment," writes Michael Blakemore, the Tony Award–winning director of Copenhagen:

> You go into a rehearsal room, which is a sort of atom, and then a lot of these rather busy particles, the actors, do their work and circle around the nucleus of a good text. And then, when you think you're ready to be seen, you sell tickets to a lot of photons, that is, an audience, who will shine the light of their attention on what you've been up to. Then something very strange happens: the thing that you rehearsed . . . and that you have seen a hundred times is put on a stage and a thousand pairs of eyes hit it and alter it. The energy an audience brings to it, the energy of their laughter and their rapt attention, changes what is there.[7]

Using this metaphor from physics, Blakemore captures the essence of live theater and describes what distinguishes it from cinema: the unique chemical reaction between audience and performers depends entirely on the uncertainty of liveness, the unpredictability of what will happen each night, the minute permutations of each performance. Such issues are explored in depth in this study in relation to specific plays such as Arcadia, After Darwin, and Copenhagen, on the one hand, and Infinities and Mnemonic, on the other.

My aim is not to provide exhaustive and comprehensive coverage of science plays; rather, it is to provide a sense, through specific examples, of what I judge to be core science plays. While themes and issues are of course important to this study, the aspect of science plays that forms the main object of interest here is their integration of form and content. This feature, which will be shown through examples in each chapter, accounts for the book's selection of plays. I hope that readers will take each chapter as a case study in the way a particular scientific idea or field has been addressed on stage. Analyses of individual plays are meant to serve as models for further study of science and theater, as well as to provide insight into a given play. It is no coincidence that science plays display a strong correlation between form and content: their questioning stance with regard to vital moral and scientific issues is often packaged in forms that reject dramatic realism and conventional staging methods. Yet so far, while they have received tremendous attention from scientists, science plays have not been widely studied by theater scholars. Scant attention has been given to the fact that these are first and foremost *plays*. Instead, critics, audiences, readers, and scholars have pounced on the themes they deal with, especially those they often share: the implications of scientific discovery, the moral responsibility of the scientist, the role of science in our lives, and the question of whether or not scientific discovery is inherently neutral.

The most striking contribution of science plays is that the best ones successfully employ a particular scientific idea or concept as an extended theatrical metaphor. They literally *enact* the idea that they engage, a performativity that is provocative and innovative and that has occurred so consistently in science plays that it is more than just a trend or a coincidence. Outstanding examples of this merging of form and content that are discussed in subsequent chapters include Dürrenmatt's *The Physicists*, Frayn's *Copenhagen*, Edson's *Wit*, Wertenbaker's *After Darwin*, Friel's *Molly Sweeney*, Stoppard's *Arcadia*, and many others.[8] The extraordinarily thorough integration of real science into the texture of these plays is one of the defining characteristics of good science plays—successfully harnessing a theatrical language to a scientific one. Plays like *Copenhagen* and *Arcadia* succeed in part because they avoid the pitfall of sloppily appropriating precise scientific concepts for vague, general purposes. They do not tempt audiences to a reductive and vague application of the uncertainty principle or of chaos theory to life in general, which is just the kind of oversimplification C. P. Snow warns against in *The Two Cultures*:

> It is bizarre how very little of twentieth-century science has been assimilated into twentieth-century art. Now and then one used to find poets conscientiously using scientific expressions, and getting them wrong—there was a time when "refraction" kept cropping up in verse in a mystifying fashion, and when "polarised light" was used as though writers were under the illusion that it was a specially admirable kind of light. Of course, that isn't the way that science could be any good to art. It has got to be assimilated along with, and as part and parcel of, the whole of our mental experience, and used as naturally as the rest.[9]

The science plays under investigation in this study have been selected in part because they make compelling dramas and in part because they all integrate their scientific ideas accurately and effectively into the art form, in this case theater, and avoid oversimplification and reduction. Indeed, scientists have praised plays like *Copenhagen* for their accurate depiction of complicated science.[10]

Clearly, theater, performance, and science have natural affinities. As theater scholar Jane R. Goodall has noted in her recent book *Performance and Evolution in the Age of Darwin*, "The equivalence of monkeys and humans had been a commonplace joke in street performances and fairgrounds for centuries" before Darwin. The theater, as well as nondramatic spectacles such as acrobatics, clowning acts, and monster shows, was a testing ground for ideas about evolution from the 1830s onward, providing a site for "popular speculation" about this scientific idea.[11] Scholars have also noted the theatrical aspect of the operating theater throughout history, captured in paintings by Eakins and others. All these modes of performance have linked theater with science and made the stage a natural site of speculation about the responsibility and motives of scientists in their acquisition and use of knowledge, and of the delicate tension between knowledge and power.

With the advent of more recent science plays that form the subject of the bulk of the chapters in this study comes a deliberate, self-conscious merging of theme and form. The two properties are no longer divorced but are interdependent; the "what" of the play is directly related to "how" it is told. Most of the recent science plays discussed in this book share this feature, which comes from the science playwright's special problem of exposition: it is not just about getting a story across, but presenting a set of ideas that can be quite complicated to explain. In some cases, like *Infinities*, there isn't even a story; the playwright and director deliberately want the audience and especially the actors to be liberated from the traditional, comforting pillars of theater such as plot and characterization. The abandonment of linear, causal structure gives this play about infinity an open-endedness that mirrors its subject matter.[12] Theater practitioners have tended to recognize and indeed seize on this characteristic of science plays. As the Danish director of *Copenhagen* said to Michael Frayn in the first of many symposia on the play, "I'm very excited about the way form and content fit together in your play."[13] Other directors whose work engages the performative aspect of so many science plays include Luca Ronconi, Michael Blakemore, and Jean-François Peyret. Peyret in particular diverges sharply from the kind of play he feels *Copenhagen* represents, and the book will explore why he does so.

The book makes a clear distinction between science and fantasy, a distinction that is often blurred or ignored in popular notions of what scientists do. Science fiction and fantasy plays lie beyond the scope of the current study and deserve more critical attention.[14] For the sake of continuity and space, the focus of this book is on real science, and on the ways in which it interacts with literature. The tendency toward fantasy and away from "hard" science characterizes the way in which science is dealt with in other media like film and television. In a sense, my method- ology of trying to merge the two discourses—scientific and literary—is a reflection of what the plays themselves do so well in bringing science to the stage. The book will ultimately show how playwrights are attempting to fuse the "two cultures" (and perhaps help dissolve C. P. Snow's famous dichotomy altogether) and to reflect the centrality of science to our every- day lives.

Science on Stage is designed as a companion to teachers, students, and general readers and audience members—scientists and humanists alike. Ideally, it will help teachers design and implement courses on science plays. It will help students understand the plays they will encounter in such courses. It will give general readers from both the sciences and the humanities a sense of the tradition of science plays and the unique achievements of playwrights of this genre through selected, outstanding, representative works. Its scholarly aim is to contribute to the field of the- ater and performance studies the first full-length investigation of science plays within the context of performance (their dramaturgical strategies and their production and reception) when most of the attention that has been paid to them has focused almost exclusively on themes and content.

While theater has increasingly concerned itself with science, far sur- passing film as the art form that most consistently and seriously engages scientific subjects, critics in the fields of theater studies, modern drama, and performance have been slower to examine this development in any sustained way. In the small body of literature on science plays, only a handful of articles deals with them from a theatrical standpoint,[15] and two books address a range of mostly thematic issues: Charles A. Carpenter's Dramatists and the Bomb, which focuses on plays about physics by Ameri- can and British playwrights between 1945 and 1964, and Allen E. Hye's The Moral Dilemma of the Scientist in Modern Drama, addressing selected plays about science, mostly German, from the nineteenth and twentieth centuries. In addition to these valuable studies we also have the broader scope of Roslynn D. Haynes's From Faust to Strangelove: Representations of the Scientist in Western Literature, which includes all literary genres, not just theater. In general, the content of science plays has been discussed exhaustively, but scant attention has been paid to their theatricality. No in-depth analysis of science plays as vehicles for the stage exists, and that is the gap the current study is intended to fill.

The book seeks to explain why science is figuring so prominently in the theater more than in other genres—why drama lends itself particularly well to the staging of science. In the course of his lecture on *The Two Cultures,* C. P. Snow noted that "the only writer of world class who seems to have had an understanding of the industrial revolution was Ibsen in his old age: and there wasn't much that old man didn't understand."[16] It is significant that Snow singles out a playwright for such praise in a discussion of the gulf between humanists and scientists. Decades later, a new wave of science plays has brought theater and science together in new and provocative ways, challenging the very notion of a gulf between the sciences and the humanities through mutual illumination. Watching plays like *Copenhagen,* we learn about physics and the history of science, and we also learn about the unique possibilities of the stage. Despite—or because of—its minimalist approach to staging, in its long run on Broadway and in the West End, *Copenhagen* routinely garnered such praise as "An evening with Michael Frayn's dazzling new drama will be amongst the most exhilarating, challenging and involving two and a half hours you ever spend in a theater."[17] The success of plays like *Copenhagen,* which seem to dispense with the usual devices of theater, prompts many questions about this genre. Science plays tend to be heavily discussion based and demand close attention from the audience. They also assume a certain level of education on the part of viewers. Such features have become standard in plays about science, and their popularity continues to grow steadily. It may be that the very departure from typical features of the drama, and the greater intellectual demands, actually account for their success. If so, what in turn does this tell us about our expectations of the audience; have we underestimated what they can and will see?

Greater public awareness of science may be desirable, but theater could also do with such enhancement of its profile; not just science but theater too may benefit from the success of science on stage. As British theater scholar Baz Kershaw has argued, "Theater has become a marginal commodity in the capitalist cultural market-place" and is in the process of being supplanted by performance.[18] These days, at least in Western societies, we are not used to thinking of theater as a politically influential art form or as a medium capable of shaping cultural practices and social attitudes. Yet for centuries, and until fairly recently, it was, as state censorship of the drama in many countries attests, most famously in Britain from 1737 until 1968. In America, a different kind of censorship was brought to bear, for example, on Flanagan Davis's Federal Theater Project (FTP), a groundbreaking organization that reached unprecedented numbers of viewers and tackled sensitive political issues head-on. It was quashed by the House Un-American Activities Committee, cut short in its prime yet with a remarkable record of achievement through its numerous theatrical productions. Chapter 3 on early physics plays deals extensively with the FTP, and it is useful to recall this once-vital and central role of theater as we embark on an exploration of the intersection of science and the stage. While the social and political influence of theater may have waned, science has risen to wield tremendous impact in terms of its findings and their implications. The ethical considerations over the uses of scientific and medical discoveries are constantly in the public eye. Consequently, the bringing of scientific issues and controversies into theaters in new and intense ways over the last decade or so has restored to the stage much of its former status as an influential art form.

Dramatists and novelists alike are finding science a mine of new ideas and insights into our behavior. The British literary critic John Carey suggests that "it is partly the result of the success of popular-science writing. . . . Novelists detect a public interest. It also reflects an awareness that science is now the dominant realm of knowledge, so novelists do not want to be, or seem to be, shut out of it."[19] Literary critic and novelist David Lodge has suggested that the novel itself as a form holds the key to the mystery of human consciousness. An investigation into the role of science in novels is much needed but lies beyond the scope of this study.

C. P. Snow suggested that "the clashing point of two subjects, two disciplines, two cultures—of two galaxies, so far as that goes—ought to produce creative chances. In the history of mental activity that has been where some of the break-throughs came."[20] As we shall see, the marriage of the resources of the stage and the ideas and issues of science does indeed bring about unprecented creative chances, and audiences seem hungry for them. The media have also caught on to this wave. Newspapers and magazines have increasingly featured science plays in their pages. Radio has taken an interest as well: in America, National Public Radio's *Talk of the Nation: Science Friday* has devoted three separate shows to science and theater, in October 2000, February 2002, and January 2004, respectively, all featuring playwrights and scientists discussing specific examples of this new phenomenon.[21] Television and film have been slower to showcase science and drama. Perhaps this is understandable given their very different genres, but in September 2002 a film version of *Copenhagen* (starring Stephen Rea as Niels Bohr) aired on TV in the United Kingdom; this was preceded several years ago by an HBO film version of Margaret Edson's *Wit* (directed by Mike Nichols and starring Emma Thompson, Eileen Atkins, and Christopher Lloyd). A film of David Auburn's *Proof* has been made, starring Anthony Hopkins and Gwyneth Paltrow.

Also instrumental in the proliferation of science plays is the establishment of international funding initiatives that encourage and support the writing of new stage pieces that engage science. The Sloan Foundation in New York sponsors a festival of new plays about science, and the Wellcome Trust in Great Britain has a similar funding program to help generate plays and films about scientific subjects. The Sigma Tau Foundation in Italy is largely responsible for the development of John Barrow's play *Infinities*. Other science and arts initiatives proliferate. The Arts Council England and CNAP, a bioscience research center at the University of York, have cosponsored a conference, "Rules of Engagement," in 2005 that featured performances of science and arts collaborations. The importance of such financial support to would-be science playwrights cannot be underestimated; economics is a major factor in the development of this genre of plays, and not surprisingly the money comes mainly from scientific organizations that are committed to raising public awareness of science.

In this spirit of public awareness, science plays have also generated widespread publicity through performance-linked symposia on specific plays.[22] Experts in science, history, and theater explain the contents of the particular play and engage in public discussions about it that often are transcribed and posted on the Web. Wherever *Copenhagen* has been produced, for instance, a symposium has occurred to complement the production. The symposium linked to a particular production, along with the pre- or postperformance "talk-back" with audience members, is not a new development in theater history—predecessors include the preperformance lecture in France in the nineteenth century that would "explain" the play to the audience, usually featuring some prominent expert.[23] In a sense, modern science plays are generating a revival of this tradition of the preperformance lecture. Often these events themselves have become highly dramatic affairs as experts with different interpretations of the play clash. These public discussions have helped to give science plays a high profile.

As these symposia have shown, the use of history and real characters, and the degree to which the playwright gets the science "right," are thorny issues in the discussion and reception of science plays. They will be extensively addressed in chapter 8. *Science on Stage* also makes a clear distinction between those science plays whose main purpose is didactic—to educate people about the particular concept or field being dramatized—and those that stage science in an aesthetically integrated way. There are many science plays whose scientific content is unassailable but whose theatricality is weak. They may teach science, but they do not make superb or even satisfying drama. The present study aims to reclaim the literary-critical currency of the term "science plays" so that we understand these works first and foremost as viable dramas, scripts for performance, literature for the stage—not merely vehicles for the teaching of science.

If conflict is the heart of any dramatic story, it is no surprise that playwrights find so much rich material in the history of science, as well as so many fictional possibilities. Certainly it is no coincidence that science plays have opened up whole new territories of subject matter for playwrights to address, beyond the stale and melodramatic material of dysfunctional families that the theater scholar Martin Esslin complained of when he noted "the narrowness of subject matter" in contemporary American drama; "much of today's playwriting for the stage can be seen as merely a continuation of soap opera on a slightly more elevated plane."[24] Although his complaint is specifically about the lack of engagement with political ideas in American drama (as opposed to the dominance of political playwriting in England, for example), Esslin points to a lack of intellectual content in many contemporary plays that continue to be solipsistically focused on dysfunctional and petty families. Science plays not only supply the missing substance; in many cases they bring politics back to the stage, since the political implications of new scientific ideas and discoveries are a central theme.

Although we are very exposed to science and medicine through the media, much of science and what scientists do remains a mystery. As the pioneering literature and science scholar Gillian Beer puts it, "The sealed laboratory lies at the centre of social fantasy. What goes on there? Do we wish to know? Are we responsible for it? . . . Such questions have dogged our culture and writing over the past 200 years, presenting themselves often in a positive form for the Victorians, more often as dread in this century."[25] Beer identifies "the mathematicization of scientific knowledge" as the reason for this change, accelerating scientific communication "to a startling degree, as if the Tower of Babel had been built in a day once the workers found a common discourse."[26]

But even more significant seems to be the need to revisit our history. So many science plays dramatize events that profoundly shaped our lives, from Galileo's discoveries and recantation to the race to develop the atomic bomb. They delve into the genesis of these events as well as their repercussions, and the bridge to the difficult science is our common humanity and the need to understand our history, regardless of which of the two cultures we inhabit. As the mathematician says of the literary critic in Tom Stoppard's play *Arcadia*: "He's not against penicillin, and he knows I'm not against poetry."[27]

At about the same time as Snow, the pioneering British theater scholar Glynne Wickham delivered his own thoughts on the "two cultures," and it is worth quoting substantially from his lectures here. Wickham concluded that "arts and sciences no longer . . . talk the same language" and that "the scientist or engineer in his concern with man's material environment is dangerously careless in his lack of concern for man's human condition."[28] Wickham noted that arts graduates know even less about science than science students do about art: "Sir Francis Bacon once boasted that all knowledge was his profession; and indeed the range of his mental activities has amazed scientists as well as students of letters ever since. Yet, however much may have been added to the world's stockpile of knowledge in the four hundred years since his birth, few scientists today have any knowledge of how to prevent the fruits of specifically scientific discoveries being used to plunge mankind into an unparalleled chaos of suffering, destruction, and new-barbarism."[29] Wickham envisions uniting the two cultures through drama: "Today, for better or worse, scientific knowledge has given us a world where what is single and unified in nature has been arbitrarily split asunder."[30] Drama has an "integrating power, a subject which can relate the ancient world to the present day, which can bring critical appraisal into direct contact with creative experiment, which can provide the arts man with a lively introduction to scientific thinking and the scientist with as lively a reflection of his own human condition."[31] Where Snow's lectures worry and warn about the dangers of the two-cultures divide, Wickham's articulate a much more positive vision and offer a solution—uniting the two cultures through drama. From Snow and Wickham to the present, scientists and artists alike have been calling for greater mutual understanding and a fruitful intersection of their respective endeavors, and science plays on the whole represent a striking way of answering that plea.

The next two chapters briefly survey the terrain of science and the stage over the last four centuries to lay out the parameters of their interaction, give an indication of which plays have dominated and indeed shaped the tradition of science plays, and see what role the science has played within the given theatrical work. Having established the rich and varied tradition of science interacting with the theater, we can go on to look at some possible explanations for its continued prominence and the appeal of science plays today. This, in turn, will provide the necessary framework for our exploration of specific contemporary science plays in subsequent chapters.

1

The Tradition
of Science Plays
•

In her book *From Faust to Strangelove*, Roslynn Haynes charts the major trends in the portrayal of the scientist in Western literature. She not only taxonomizes the many different manifestations of the scientist figure but also shows how such depictions directly or indirectly reflect societal perceptions and cultural beliefs about science and its professors, intricately linking science, literature, and culture. While her study does consider selected dramas, its main focus is on novels, short stories, and other fictive genres. Yet it is strikingly clear that within the long history of science in literature, the theater has been one of the most consistently prominent sites of engagement between the two cultures.

Although not an exhaustive survey, this chapter provides a taste of the way science and theater have intersected over the centuries and sets out the parameters for the tradition of science in the theater: not just identifying the representative plays but discussing how the science figures in them. Broadly speaking, science moves from the margins to center stage in the drama—from simile and metaphor to thorough structural and thematic integration. Setting out this development and furnishing a framework for the plays that are analyzed in some depth in later chapters will enable us to contextualize them better and to see not only how they relate to one another but how they correspond to the tradition and its trajectory. Plays that engage medical themes and motifs and/or depict doctor-patient relationships are also part of this tradition but will be considered in a separate chapter.

From Metaphor to Performance

The way in which scientific themes and ideas have been engaged by playwrights and directors has become increasingly sophisticated. In Elizabethan examples, like Shakespeare's *Coriolanus* and Marlowe's *Doctor Faustus*, scientific and medical themes can provide handy illustrative metaphors. In the former, the metaphor of disease richly illustrates the political problems of the main character, as the corrupt state is likened to a body with diseased limbs in need of amputation: "He's a disease that must be cut away."[1] Here, the science is used to make a metaphor that describes the situation or character. In general, earlier plays that deal with science do so in this literary fashion—borrowing images or ideas to make metaphors. While Coriolanus is likened to a diseased body, the common people are a "herd of—Boils and plagues," on whom should descend "all the contagion of the south . . . and one infect another."[2] The metaphor of disease is vividly used, but the form of the play is divorced from its content. In *Doctor Faustus*, science is incorporated into the play in a very different way: science itself, or the scientist, becomes the embodiment of the problem, not just a metaphor for it. The problem of human overreaching, of trying to know too much, is personified in the figure of Faustus.

From these two Elizabethan examples to the present day is a long leap, and a significant shift of emphasis occurs between earlier and later science plays in terms of how the science figures in the play. Later science plays—the ones that define the new wave of science playwriting in the 1990s and beyond—move toward a formal and structural integration of the science. The playwright structures his or her play according to the scientific idea at its core, making the piece performative in nature, as outlined in the introduction.

From *Theatrum Mundi* to the Stage as Cosmos

The time-honored cliché of the stage as a microcosm of the world is perhaps best known from Jacques's soliloquy in *As You Like It*, which begins:

> All the world's a stage,
> And all the men and women merely players;
> They have their exits and their entrances,
> And one man in his time plays many parts,
> His acts being seven ages.[3]

This was a favorite image of Shakespeare's; for example, in the prologue to *Henry V* he conjures "a kingdom for a stage," beseeching the audience to use its imagination and think of "this wooden O" (the circular Globe Theater) as the world, housing "vasty fields," the perilous ocean, "mighty monarchies," and advancing armies "within the girdle of these walls."[4] Shakespeare may have made it globally famous, but this trope of stage as world "can be traced back to the Middle Ages and beyond, to Plato—who already disliked the institution from which he derived the image."[5]

By the twentieth century such imagery is well established in plays and firmly links the theater and the universe in the audience's imagination. In the mid–twentieth century, Brecht has Galileo use the stage to show Andrea how the Copernican theory works in the opening scenes of *Galileo*, with an iron washstand as the sun and Andrea on a chair as the earth.[6] Thornton Wilder's *The Skin of Our Teeth* (1942) and Ewan MacColl's *Uranium 235* (1947–52) both in different but highly visual ways use the stage-as-world device to transform the stage into a living cosmological symbol. More recently, Stoppard's *Galileo*—only recently published—and *God and Stephen Hawking* by Robin Hawdon posit the stage as the cosmos in miniature. Some of these examples will be discussed in greater depth later.

The Scientist Sells His Soul

The ur–science play may well be *Doctor Faustus* in terms of both text and performance. As Mark Berninger points out, the "long tradition of dealing with science in the theater . . . goes back at least to the various dramatisations of the Faust myth."[7] Marlowe's *Doctor Faustus* helped (literally) set the stage for the science play by establishing literary roots and a theatrical model for the integration of science and theater. Between it and modern science plays lies some important common ground, but also some key distinctions and divergences.

Faustus's bargain with the Devil sounds an ominous note as one of the first dramatic representations of the scientist; this negative image signals a distrust of science that becomes one of the defining characteristics of science plays for centuries. If "the relationship between science and drama has been predominantly one of opposition," the tension surely begins here.[8] Marlowe's Faustus is not a passive Everyman caught in the inexorable wheels of fate; he is solely responsible for his own demise, and yet in theory his desire for greater knowledge is admirable. The dilemma posed by this tension between wanting more knowledge and flying in the face of religious doctrine finds its way into later science plays, like *After Darwin*, *Galileo*, and *Inherit the Wind*.

Doctor Faustus's central concern with the pursuit of knowledge and the use and abuse of that knowledge makes it the archetypal science play, and this concern only gets stronger as the issues of modern science become ever more ethically complex. In particular, the theme of intellectual curiosity leading to Icarus-like overreaching recurs throughout the science play canon, but through the opposite idea of "underreaching." Put simply, modern science plays, like Stephenson's *An Experiment with an Air-Pump* and Brenton's *The Genius*, often revolve around the idea of science having gone too far and created a hell on earth. They invoke the hell motifs of *Doctor Faustus* to convey the enormity of the implications of postwar science.

Most important, Marlowe foregrounds the illusionistic practices of the stage as he shows what comes of Faustus's supreme knowledge: the ability to perform cheap tricks and sleights of hand that are more closely linked to the artifice of the stage than to the practice of science. The play makes an implicit connection between theatrical illusion and sham science. "Then in this show let me an actor be," says Faustus in act 3 as he and Mephistopheles are devising their treatment of the pope, one of several metatheatrical moments in the play when Faustus acknowledges himself as a performer. Faustus's daring and blasphemy do not get him the great power, riches, and unsurpassed knowledge that he envisions in the play's opening scenes; instead, he becomes a theatrical impresario, able to conjure spirits (and grapes) and to play pranks on the pope. In this overt analogy to theater itself, which similarly pulls the wool over people's eyes through artifice, it is as if Marlowe is cautioning that such longing for superhuman power leads only to shoddy theatrical tricks. All the while, of course, he is paradoxically validating theater as the site of the exploration of serious issues.

This "conscious theatricality"—Richard Allen Cave's term to describe the strategy of Marlowe's fellow dramatist Ben Jonson in *The Alchemist*[9]—anticipates the more overt performativity of contemporary science plays. Right from the start, science plays are conscious of the way form can enhance and convey theme, and surely the "liveness" that is unique to the theater plays a tremendous role here. From Marlowe to Thornton Wilder in *The Skin of Our Teeth* to the contemporary examples that are the main focus of this study, playwrights have used science specifically to probe the problem of human progress through the advancement of knowledge.

Not everyone thinks highly of *Doctor Faustus*. "What can be said in defence of *Faustus*?" despairs one reviewer of a recent revival. "Has anything else so ramshackle ever entered the Pantheon of masterpieces?"[10]

What some scholars claim is Marlowe's "supreme achievement," "a kind of enchantment,"[11] others find his "most famous if least effective play," which reads and acts like

> a sequence of Monty Python sketches topped and tailed by a few speeches with flecks of poetry in them. The Germans should subsidize Marlowe for putting Goethe in so good a light. . . . *Doctor Faustus* is little more than a vulgarized morality play into which a young doubter and man-about-the-university has inserted some shallow notions of atheism. It lacks even the courage of its conventional defiance, always veering off into stereotyped repentance.[12]

The influence of the old morality plays, "whose remnants are ruinous in *Faustus*," along with the "clumsy kind of magic that fills" it, makes it poor drama.[13] Such views press the question of whether the play is as good as what it represents. Perhaps the play's literary and theatrical merits have been overblown, and its significance and continued relevance really rest with its status as precursor of "science plays."

At first glance, *Doctor Faustus* hardly seems like a "science play" because the science in the play is limited to a few passages about astronomy, and anyway the whole notion of what science was at that time is fundamentally different from how we understand it today.[14] But one can argue that it is not the quantity of the science in a science play that matters, but the quality of its integration: the way in which it figures both thematically and theatrically. A second point relates to the nature of scientific discovery. Much of science is highly progressive and therefore potentially subversive of the status quo, destabilizing norms by constantly questioning and changing what we know. By putting the archetypal scientist on the stage, Marlowe helped to establish a link between theatricality, science, and subversion that we find again and again in science plays. What the editors of the Revels edition of *Doctor Faustus* call the "subversive energy released by the theatrical experience of this play" helps to explain why so many writers have chosen theater as the medium to explore subversive ideas coming from science, since historically theater and science share this subversive, dangerous, and carnivalesque quality.[15] It is a natural fit, then, to bring them together.

As J. W. Smeed points out in his book *Faust in Literature*, two major changes happened in post-Marlowe depictions of Faustus. First, the idea of an *illicit* knowledge, that is, forbidden black magic and necromancy, gives way to a far more positive, post-Enlightenment notion of knowledge not for the betterment of the individual but for the improvement of society. The leap from Faustus's declarations of his ambitions to rule the earth (linking knowledge with power of a personal, ambitious sort) to our concerns with the impact of scientific discovery on the common good is great and significant, casting the scientist in a far more positive light by stressing intellectual curiosity rather than overreaching hubris. Postwar science plays, beginning with the final version of Brecht's *Galileo*,[16] have become much more ambivalent about the motives of the scientist.

The second major change in the depictions of Faust, or at least his indirect descendants in literature over the centuries, is the largely Germanic-inspired emphasis on romance in the story, to the subjugation of science. An example of this is in Gounod's opera, with its reductive focus on the doomed love affair between Marguerite and Faust. What is striking is that modern science plays, with few exceptions, have followed Marlowe's example rather than Goethe's, focusing on the themes relating to science and its pursuit much more than love interests. In this way, science has remained at the core of contemporary plays about science. Love, when it exists at all, is secondary or tangential, for reasons outlined in chapter 2.

It remains to be asked: Would we have had a tradition of science plays without *Doctor Faustus*? Probably. But this play arguably sets up a paradigm of using theater as a place to unite the two cultures, by raising fundamental questions about the nature and ethics of scientific pursuit, the personality of the scientists involved, and the connection between the aspirations of the scientist and the theatrical terms in which such aspirations are framed.

Pseuds' Corner: From Jonson to Ibsen and Shaw

Ben Jonson's *The Alchemist* features overreaching scientists of a very different character—fakes, pseudoscientists, and poseurs bent on exploiting the mysterious science of alchemy for financial gain, not for the sake of further knowledge. In *The Alchemist*, three sham scientists launch a dubious scheme they call the "venter tripartite," in which Face, Dol, and Subtle pose as an alchemist and his assistants and entertain a steady stream of customers eager to buy what they believe the alchemists can purvey, from sexual prowess to wealth to professional success.

The play deserves inclusion in this discussion because the allure of alchemy lurks around the edges of the science play tradition, whether in the form of actual alchemists, as in Jonson's play, or in more suggestive mode, as in Tony Harrison's theater piece *Square Rounds*, in which the real-life figure of chemist Fritz Haber furnishes an alchemical motif. As August Strindberg's "inferno" crisis illustrates (the period in the 1890s when he wrote nothing but papers about his often dangerous alchemical experiments), alchemy has continued to fascinate artists and scientists alike. There is something poignant as well as sinister in the alchemical enterprise, something fundamentally human, and playwrights from Jonson onward have used that dualism to great effect. And, as Gillian Beer has noted, alchemy is specifically linked to the theme of transformation, and to the theater's power to effect change.[17]

Jonson's satirical target is not the alchemists but their customers, the willing dupes whose vanity and greed drive them to seek the secret knowledge of the would-be alchemists. The kind of overreaching these gulls exemplify is, by implication, far worse than the rather imaginative and savvy sort displayed by the cozening alchemists. At least these alchemists have done their research. Jonson is thoroughly conversant with the technical jargon of alchemy. It is remarkable how in this polemical play—just as in Shaw's *The Doctor's Dilemma* three centuries later—even while the author inveighs against scientific fakes who dupe their unsuspecting followers, he shows a real grasp of the field that furnishes his dramatic material. As the sly alchemist Subtle quizzes his accomplice Face, Jonson's dialogue reflects—and has great fun with—his familiarity with Renaissance science:

> SUBTLE: Name the vexations, and the martyrisations
>   Of metals, in the work.
> FACE: Sir, putrefaction,
>   Solution, ablution, sublimation,
>   Cohobation, calcination, ceration and
>   Fixation.
> SUBTLE: This is heathen Greek, to you, now?
>   And when comes vivifacation?
> FACE: After mortification.
> SUBTLE: What's cohobation?
> FACE: 'Tis the pouring on
>   Your *aqua regis*, and then drawing him off,
>   To the trine circle of the seven spheres.
> SUBTLE: What's the proper passion of metals?
> FACE: Malleation.

SUBTLE: What's your *ultimum supplicium auri?*
FACE: Antimonium.
SUBTLE: This 's heathen Greek, to you? And, what's your mercury?
FACE: A very fugitive, he will be gone, sir.
SUBTLE: How know you him?
FACE: By his viscosity
   His oleosity, and his suscitability.[18]

Far from being gibberish, such language shrewdly satirizes "a broad range of popular superstitions and pseudo-sciences," and it succeeds because Jonson knows the subject of his satire extremely well. He also knows his audience: "Queen Elizabeth and several nobles invested in alchemical projects, and its possibilities continued to fascinate learned men into the late seventeenth century." Jonson knew the work of the mathematician John Dee, a possible model for Subtle, who also "practised alchemy and attempted to communicate with spirits."[19]

Given that Jonson's target was the audience and not the alchemists, it is intriguing that the play was a "resounding popular and artistic success . . . one of the few Jonson plays which seems to have caused no trouble of any kind for its author."[20] Perhaps he was simply very adept at flattering his audiences into thinking of themselves as the clever ones rather than the gulls. Perhaps the way he depicted the gulls made the audience want to think of themselves as the smarter, albeit unscrupulous, alchemists rather than their customers. Whatever the reason, Jonson succeeded in disguising some fairly harsh criticism of his audience, a subversive strategy rather akin to Oscar Wilde's similar success in marshaling the tools of Victorian theater to attack his audience's hypocrisy. The fact that *The Alchemist* was so well received by its audiences indicates how subtly Jonson's satire works.

In addition, the play offers a comical commentary on theater itself, complementing Marlowe's more sobering analogy. *The Alchemist* presents "a series of farcical characters each intent on realising a *better* self come in the clutches of a trio, consummate actors all, who in a mercantile age are choosing to sell illusion as a commodity. With the aid of a few tawdry props and appropriate foot-and-half-foot words they give their clients the chance to star in their various private scenarios—at a price, of course."[21] The "tawdry props" are not so different from those employed by Mephistopheles and Faustus, with a similar purpose: to deceive and trick. As Cave points out, alchemy and theater coincide:

Jonson draws and sustains throughout the play a brilliant conceit relating alchemy, the refining of base metals into purest gold, with acting (the transformation of humble mortals into heroes). Subtle's alchemical laboratory we never see (patently it does not even exist); instead we watch the effects of the trio simply talking to their clients about its potential. All the magic is in the power of their words to feed each client's imagination until it invents a brave new world to its own satisfaction. By making these moments of self-transcendence plays within the larger play-structure, each carefully staged and paced by Face or Subtle, Jonson cunningly transforms a superb social comedy about Jacobean London into a serious disquisition on the nature of theater.[22]

Paradoxically, the power the three rogues wield lies in the suggestion of alchemy rather than its actual demonstration—their laboratory remains off stage, behind a curtain, forever mysterious and impenetrable.

We have seen that what Cave refers to as the "conscious theatricality" of the play applies equally to *Doctor Faustus*, to *Galileo* and other midcentury science plays, and to recent plays like *Copenhagen*, *Wit*, *After Darwin*, and *Arcadia*. Some scholars have found a connection between Jonson and Brecht in their respective use of metatheater; Cave, for instance, has argued that Jonson's "wholly purposeful theatricality" is concerned with creating "an 'alienated' audience, one made conscious (as Brecht would have his spectators) of participating in the shaping of an artifice the better to perceive the imaginative consequences signified by the performance."[23] In fact, one of the hallmarks of science plays, especially the most recent ones, is precisely this attribute, as analysis of selected plays in subsequent chapters will show. Brecht's imprimatur is deep. In the tradition of science plays, the works that seem least effective are those that adhere most closely to a formulaic dramaturgy such as the well-made play (Brieux's *Les Avariés* is an example), while the ones that stretch and prod the norms of accepted theatrical modes have been most enduring and successful.

It should be noted that scientists and their laboratories crop up frequently in dramas throughout the Restoration and eighteenth century, mainly to provide easy targets for ridicule and fodder for satire, lampoons, and spoofs. As is well known, Thomas Shadwell's *The Virtuoso* (1676) lampoons the Royal Society and specifically the famously diminutive scientist Hooke.[24] The motif of the doctor or scientist as fossil collector, dry and out of touch with reality and closeted in his museum of useless curiosities, appears in plays like *Three Hours after Marriage* by "Scriblerus"

(John Gay, 1717). The former features Dr. Fossile, a charmless predecessor to Shaw's Dr. Walpole, who has one standard cure for any ailment: where Walpole unfailingly diagnoses blood poisoning and prescribes the surgical removal of the mythical "nuciform sac," Dr. Fossile repeatedly advises dosing his patients with "a Quieting Draught."[25] Science heavily influences this play; the dialogue is larded with medical and scientific jargon designed to raise a laugh, and the setting is a spectacularly cluttered "museum" or laboratory full of relics and curiosities collected by Fossile. *Three Hours after Marriage* sends up the notion of the intellectual in several scenes that show vain and smug scientists engaged in vigorous one-up-manship, surrounded by the useless curiosities Fossile has collected in his research and that now cram his cabinet—including an alligator. When it figures at all, science tends to be the subject of ridicule in the drama of this period, an interesting social commentary on its role within culture.

Brecht, Galileo, and the "Conscious Theatricality" of the Science Play

Any discussion of science plays must include quite centrally Brecht's *Life of Galileo* (referred to hereafter as *Galileo*). Although there have been other dramatizations of the figure of Galileo, Brecht's is the best-known play about the great scientist. It is also highly controversial, a thorn in the side of historians and scientists who believe Brecht muddles history, biography, and science in a dangerously misleading way for political ends. I will address the play with regard to these ongoing questions about how Brecht used science and history, and consider why the play remains a watershed in the development of "science plays."

Brecht wrote several versions of *Galileo* over two decades, and they differ radically with regard to the depiction of the scientist. In his now famous introduction to the play, Eric Bentley—Brecht's English translator and champion, who did for Brecht in the English-speaking world what William Archer had done for Ibsen—summarizes how the drafts of the play changed and how the versions differ from one another. He notes that the play is remarkable in that it exists in "two broadly different forms," the version of 1938 and the version of 1947. The dates alone would suggest what happened between the two versions: World War II and the use of the atomic bombs on Hiroshima and Nagasaki. It is perhaps no surprise that Brecht's 1938 Galileo is a "winning rogue" who could sabotage and subvert the dominant regime (a potent analogy with the resistance to Hitler), while the 1947 version was designed "to make the audience dislike him."[26] Brecht himself put it thus: "The atomic age made its debut . . . in the middle of our work. Overnight the biography of the founder of the new system of physics read differently."[27] The point in the postwar version of the play—the one that is most widely read and performed today—was "no longer to demand from the authorities liberty to teach all things but to demand from the scientists themselves a sense of social responsibility, a sense of identification with the destiny, not of other scientists only, but of people at large. The point was now to dissent from those who see scientific advance as 'an end in itself,' thus playing into the hands of those who happen to be in power."[28]

This shift in the portrayal of the scientist is partly what makes *Galileo* a watershed science play. Generally, from hereon in, science plays either directly or indirectly engage this notion of the social responsibility of the scientist. The scientist is rarely the underdog anymore. In addition, *Galileo* broaches a host of important questions about the nature and pursuit of science, especially its relationship to society. It shares with *Doctor Faustus* an engagement with the themes of the scientist as overreacher, the clash of religion versus science, the scientist as hero or villain, and the application of scientific knowledge. Nowhere is this captured better than in Brecht's depiction of Galileo as a sensual, earthy man who loves his food and wine and needs money to support these basic needs. To this end, in Brecht's play, he is not above "modifying" (plagiarizing) the newly invented telescope and claiming it as his own work. The scene in which he unveils his discovery in public, performing before an admiring crowd, echoes the cheap tricks of Faustus aided by Mephistopheles. Where Faustus has Mephistopheles to help him stage the show, Galileo has his actual scientific knowledge.

Another theme Brecht's *Galileo* introduces and that will become standard in later science plays is the tension between the desires of the individual and responsibility to the community. This is, of course, related to the theme of religion versus science. In *Doctor Faustus* and *Galileo*, the protagonists test their individuality against the dictates of the church to think of God and the community before the self. In both plays there is also the implication (much more outspoken in Brecht) that the church has a lot invested in keeping people in the dark, although Marlowe clearly seems to be more on the side of the church than does Brecht. But for both, the individual's search for truth is problematic; Faustus and Galileo are lone fighters for greater knowledge who ultimately isolate themselves. Although Galileo goes on to publish even more important findings than those he had to recant, he is all alone and in jail, blind, his health ruined, with few friends still in touch with him. Brecht wants to show the importance of social responsibility and the emphasis on the community rather than the individual. A similar impulse underpins *Doctor Faustus*: when Faustus deserts his community of scholars, they show concern, but he is already forging his pact with the Devil. It is as if to say, once the individual becomes more important than the group, he or she is lost.

Many other versions of Galileo's story have been written for the stage. In the 1940s, the air was thick with Galileo plays: the New York agent Audrey Wood declined to hear Charles Laughton read *Galileo* in a bid for her to produce it in New York because "she also represented a client named Barrie Stavis who had written a Galileo play entitled *Lamp at Midnight*."[29] While Brecht was working on the English version of his play with Laughton, Morton Wurtele, a scientist he had employed to advise on the scientific aspects of the play, came across a five-act tragedy about the life of Galileo written by the German playwright Arthur Trebitsch. Entitled *Galileo Galilei. Ein Trauerspiel in fünf Akten*, it was written entirely in blank verse and was published in Berlin in 1920.[30] This discovery did not seem to faze Brecht. Frank Zwillinger's 1953 play about Galileo depicts a happy ending: "The archbishop of Sienna assures him that, owing to a technicality, the Inquisition's ban on the teaching of the Copernican system is invalid, and hence such teaching is not heretical."[31]

In its fifty years of publication and production, Brecht's *Life of Galileo* has seen challengers, revisers, and adapters, as well as continuing commentary and analysis. In this regard, too, the play takes a special prominence in the tradition of science on stage, generating its own legacy of ongoing intertexuality. Several prominent playwrights have taken on the subject of Galileo, whether overtly responding to Brecht's depiction or not. David Hare and David Edgar have both adapted Brecht's play. Tom Stoppard wrote an unpublished version of *Galileo* (discussed later in this chapter) that was meant to be performed at the London Planetarium. In 1980 Howard Brenton wrote an English version of Brecht's *Galileo*, directed by John Dexter at the National Theater with Michael Gambon as Galileo, which Brenton calls "a production of great clarity and force."[32] Mel Gussow hails Gambon's "breakthrough performance" as Galileo in that production.[33] Philip Glass wrote an opera, *Galileo Galilei*, staged in 2001. The historian of science Lewis Wolpert has written a kind of duologue called "Good Evening, Galileo," in which Galileo is interviewed after seeing Brecht's play. Its main purpose is to criticize Brecht's version for historical inaccuracies and scientific distortion.[34]

Some of the authors of plays about Galileo simply want to write about this fascinating figure and the discoveries that changed the way we see the world. Others are bothered by the way in which Brecht used biography, science, and history, taking issue with his negative depiction of Galileo and his critical stance toward science itself. For example, Stoppard's research into Galileo's life and science led him to conclude that "Brecht's play was nonsensical in certain historical respects"; Stoppard wanted his version to be "essentially faithful to history."[35] To this end, throughout Stoppard's version a narrator tells the audience "when and how the play deviates from historical fact, thereby serving as a subtle rebuttal to the way Brecht manipulated the story to fit his own ideological ends."[36] While historians have been piqued at what they see as a misrepresentation of events, scientists especially have been quick to point out that Brecht's science is oversimplified, distorted, or simply wrong. Wolpert's Galileo says:

> Brecht shows no understanding of science. In the first scene of the play he has me gabbling on about a new age, with everyone questioning old certainties. What nonsense. Then there is that line where someone says "we shouldn't have tried what is in the books, but should have looked for ourselves." What nonsense. Any looking makes it obvious that the earth stands still and the sun moves around it each day. No, science first requires a sound knowledge of mathematics and a special mode of thought. To deal with matters scientifically it is necessary to make abstractions from the concepts of weight and speed, which are infinitely variable. I would never have prophesised [sic]—as he has me do—that astronomy—a mathematical science would be discussed in the market place. Brecht just wants to see our new science overthrowing the established political order. The common man will never make contributions to science. It goes, unfortunately, against common sense. Even Aristotle understood this clearly when he said that the effect of achieving understanding is to completely reverse our initial attitude of mind.[37]

Such protests at a playwright's use of biography and history—dramatizing actual events and people—are now fairly routine and offer a fine parallel case study to Frayn's *Copenhagen* and the storm of controversy it has met, the implications of which are addressed in chapter 8. The objections to Frayn's depiction of Heisenberg echo those about Brecht's depiction of Galileo. In the former case, critics have felt the playwright makes the character too sympathetic and endows him with moral qualms about the use of science where he had none; in the latter case, critics have found the scientist depicted not sympathetically enough. Above all it is Brecht's rather coarse Galileo himself that has irked many who find the emphasis on Galileo's very physical and sensual appetites distorts the truth and deflates the image of the great scientist. When he gives in to the threat of pain, he is simply being human. Yet many oppose this dramatic treatment, wanting Galileo instead to be upheld as a martyr, a hero for science.

In the case of Galileo, the discoveries he made helped to bring about a fundamental change in how we view the world, and Brecht tries to make this the focus of his play by deliberately withholding any dramatization of the courtroom scene that would doubtless form the centerpiece of a more traditional play about Galileo. Without the distraction of a classic courtroom scene, the play foregrounds much more sharply the paradigmatic shift Galileo's discoveries signify. Challenging or shedding old ideologies requires intense questioning of dominant discourses and institutions. The potential good of new scientific discoveries cannot be fully realized without concomitant advances or transformations in thought and perception.

In understanding Brecht's (or indeed any other playwright's) *Galileo*, the notion of paradigm shifts becomes highly relevant. Reading the play in light of Thomas Kuhn's work on the nature of scientific revolutions helps prevent the inordinate emphasis on biography that the Italian director Luca Ronconi laments (see this book's conclusion) and keeps the focus on historical forces and the progression of science. It helps to contextualize Galileo's struggles if one sees him as the representation of a new way of thinking that is up against not only entrenched and persistent modes of thought that need to be replaced or overthrown but also power bases such as the institution of the church that stand to lose a lot if the new thought is right. The conflict between science and the church in Galileo epitomizes Kuhn's argument that the only way science progresses is by one paradigm replacing another; it illustrates the inability of two conflicting paradigms to coexist.[38] We see this same idea dramatized in other science plays like *After Darwin* and *Arcadia*, although the latter cleverly subverts Kuhn's notion by having the representatives of the two clashing systems (Newtonian and Einsteinian) cohabit the stage toward the end of the play—two hundred years apart, yet all clothed in the same Regency garb. Characteristically, Stoppard employs theatricality as well as textuality to convey such a theme; a simple matter of staging instantly conveys a complex idea.

The Influence of Brecht on Science Playwriting

The themes of science clashing with religion and the scientist as over-reacher can be found most powerfully embodied in the figure of Brecht's Little Monk, who is caught between the desire to pursue science and the fear of what it will do to people like his peasant parents and their faith. He argues with Galileo about the uses of science and the aims of the scientist. Why not simply limit what you investigate? You could still pursue science but focus mainly on applied science with a direct bearing on improving people's lives through practical uses, like irrigation. Galileo retorts, "How can new machinery be evolved to domesticate the river water if we physicists are forbidden to study, discuss, and pool our findings about the greatest machinery of all, the machinery of the heavenly bodies?"[39] Is it overreaching to study these things?

These questions inform two interesting plays written in the aftermath of World War II, $E = mc^2$ by Hallie Flanagan Davis and *Uranium 235* by Ewan MacColl, that will be discussed in chapter 3. Both of these plays ask the audience to weigh the pros and cons of nuclear physics. They first educate the audience about the science in an entertaining style, then dramatically end by placing the decision about how to use nuclear energy in the audience's hands: Which path will we choose? it's up to us. Both the mode of the questioning and the question itself are deeply influenced by Brecht.

The Skin of Our Teeth, by Thornton Wilder, also bears the imprint of Brecht in terms of both theatricality and themes, and provides an example of how a play can contain very little actual scientific detail yet still engage the ramifications of science in profound ways. Wilder's deeply antiwar play, premiered during "the darkest years" of World War II, takes a romp through the history of humanity and a bleak look at scientific progress and discovery. The Thomas Edison–like scientist-inventor, Mr. Antrobus, "is busy inventing the wheel, the alphabet, and the answer to ten times ten while the rest of the world struggles to survive an ice age, a war, murders, and general social breakdown."[40] Antrobus "has nothing to offer in such dire straits; science cannot help us in times of global catastrophe."[41] Instead, we need to survive through ordinary people's instinct for survival "by the skin of our teeth," the ingenuity, "resourcefulness and endurance of the ordinary individual multiplied over the whole population. This, Wilder believes, is how the human race has hitherto escaped extinction and will perhaps continue to do so."[42]

As with so many of the other midcentury American plays under discussion in this book, such as Flanagan Davis's $E = mc^2$, Wilder's model is epic theater, and he stylizes his work accordingly. The play is highly metatheatrical and eschews identification through the device of Sabrina, the maid, whose wry, funny, and unemotional commentary, addressed directly to the audience, serves to preempt any temptation to empathy on its part. She constantly reminds the audience that she is an actress and they are in a theater and keeps us informed about how the performance is going (not well) and what is happening backstage (mayhem). She is, in a sense, the personification of Brecht's idea of *Verfremdungseffekt* (the alienation effect), by which audiences are prevented from identifying too closely with a character and are kept at an emotional distance. Alienation (or, more accurately translated, "making strange the familiar") is one of the cornerstones of Brecht's epic theater, his alternative to what he called (pejoratively) the "dramatic" or mainstream theater. Although one must be quite clear about the distinction between Wilder's agenda and Brecht's—the latter using alienation for directly political purposes—other Brechtian devices in *The Skin of Our Teeth* include an episodic structure, a plot that is epic in its scope (spanning the dawn of time to the present), constant interruptions, direct address to the audience (including invitations to them to go outside and smoke while the actors rehearse a scene), and the use of projections and other means of announcing and commenting on the action visually. It should be noted that all this adds up to a highly visual and entertaining theatrical feast, not at all textbook drama.

In the play's final act, time and philosophy are interwoven visually as a procession of philosophers comes forward, each bearing a number that corresponds to a specific time on the clock (itself a scientific construct). The conceit is that each one represents a point of our collective progression. Yet it is an incomplete clock; it has only the hours of darkness (or late morning, depending on your interpretation), with Spinoza as nine o'clock, Plato as ten, Aristotle as eleven, and the Bible (Genesis) as twelve. Whether we are to view the Word of God as the origination (in the beginning was the Word) or the culmination of human progress is not made clear. In the midst of arranging and explaining all this to the audience, an actor tells us that if all had gone according to plan, the stage would have been set up to mimic the solar system; unfortunately, the chorus that was supposed to carry this out has fallen ill, "so you'll have to imagine them singing in this scene. Saturn sings from the orchestra pit down here. The Moon is way up there. And Mars with a red lantern in his hand, stands in the aisle over there—Tz-tz-tz. It's too bad; it all makes a very fine effect."[43]

There is a similar unperformed stage-as-cosmos scene in a later play that has only recently come to light: Tom Stoppard's *Galileo*, the manuscript of which is housed in the Ransom Humanities Center, was published in 2003, with a brief introduction by the playwright.[44] It is Stoppard's only "post-Rosguil [*Rosencrantz and Guildenstern are Dead*] stage play never to be produced," and it started out as a screenplay commissioned by Paramount Pictures.[45] Stoppard writes in his introduction to the published text: "I had the brilliant idea of turning it into a play to be staged at the London Planetarium, where the audience would see what Galileo was looking at—the 'new star,' the moon's dark luminescence, Jupiter's moons, the sunspots, the famous comet. The Planetarium people kindly showed me the set-up. But the brilliant idea turned out to be hopeless. Lighting the actors, even if one could and even if they could have found room to act in under the huge projector, would have washed out the night sky. So I gave up."[46]

In his book on Stoppard's plays, the theater historian John Fleming notes that "Stoppard deliberately crafted his script for the Planetarium because it had a projector that could create various sky effects appropriate to Galileo's story."[47] One can see why, despite its technical capabilities, the Planetarium did not consider the project feasible. The play requires "twenty-seven speaking roles (fifteen of which can be doubled) and at least ten supernumeraries."[48] At one point in the play the church officially censures and prohibits discussion of Copernican theory, and for this scene of the reading of the censure "Stoppard sought a difficult stage effect: 'As the Secretary reads, the eleven theologians are gently lifted into the air in a slow arc, like the arc the earth makes round the run [*sic*], and at the same time, they gently rotate as the earth does on its journey.'"[49] The theologians then "pass out of view" after the reading is finished. The first act was supposed to end with similarly spectacular effects: "with the dome of the Planetarium showing the night sky as it appears to the naked eye and as it appears through a telescope, thereby breaking down some of the mysteries of the universe, showing things not possible in the Aristotelian conception of the heavens."[50]

It is interesting that no audience has yet seen these accomplished playwrights' cosmic visions enacted; Wilder's scene is deliberately withheld and left to the imagination as part of his Brechtian ploy, and Stoppard's has simply never been performed. Yet these scenes show both playwrights devising similar ways of making the audience see the stage as the universe.

And both are keenly aware of the importance of theatricality; their plays reject straight, unexamined realism, or the illusion of it. We have already seen this in the kinds of techniques and devices Wilder uses. Stoppard's use of theatricality would have come primarily from a single effect of his choice of venue: "A bulky machine, the Planetarium's projector is a permanent fixture in the center of the intended semicircular playing space. Thus, Stoppard felt that it would sabotage any attempt at illusionism."[51] Having the audience see a play in a planetarium, with its vast, airy space and soaring night skies, is already such a radical departure from the norms of theatergoing that one would think that the choice of venue alone would have sufficed to provide the desired degree of self-consciousness. The additional presence of the large projector would have made it impossible for the audience to lose itself in the action—modern technology constantly intruding itself on, indeed at the center of, the action of this Renaissance setting.

The Scientist as Underreacher

There is another sense in which *Galileo* serves as a watershed science play. Since *Galileo* (and the end of World War II) many science plays have explored physics and its consequences, with one theme in particular coming to the fore: the impossibility of "underreaching," or taking back knowledge once it is attained. In *Galileo*, the church is threatened by scientific advances; it is a problem of two institutions, two belief systems, clashing. Not surprisingly, ever since the bombing of Hiroshima and Nagasaki, the emphasis in science plays has shifted dramatically away from warring institutions to the survival of the planet. After *Galileo*, we see many science plays exploring the idea of limiting or indeed reversing scientific knowledge and its acquisition—underreaching rather than the old Faustian notion of overreaching.

In Friedrich Dürrenmatt's play *The Physicists* (1962), the focus of further analysis in chapter 3, an apparently insane physicist who calls himself Möbius has stumbled on the theory of everything and fears how it may be used in the wrong hands. He shuts himself up in an asylum pretending to be mad in the self-sacrificial hope that this act will "unknow" his discovery. "We have to take back our knowledge and I have taken it back," he says, explaining how he has destroyed his manuscripts—only to be outwitted by the evil doctor who runs the asylum and who has secretly copied the material before it was destroyed. In the end, the defeated scientist must admit that "what was once thought can never be unthought."[52]

A similar predicament dogs Leo Lehrer, the protagonist of Howard Brenton's play *The Genius* (to be discussed in depth in chapter 3). We may remember in *Doctor Faustus* that although Faustus frequently shows some anguish and doubt over selling his soul, he refuses several chances to repent. By contrast, both Möbius and Lehrer see that they are on the path to hell and try to repent. This seems consistent with other modern science plays, as the scientists tend to be guilt-ridden and repentant about endangering humanity through their discoveries. Yet the crux of these plays is the very notion that is central to *Doctor Faustus*—that there are limits to human knowledge, and stepping over those boundaries leads to individual and mass destruction.

Another physics play that suggests the theme of "underreaching" is Frayn's *Copenhagen;* the theme of trying to take back knowledge is prominent here, too. The scientists in the plays by Dürrenmatt and Brenton both want to unthink thoughts they have had, discoveries they have made. In *Copenhagen*, we are given three possible reasons why the German physicist Werner Heisenberg failed to lead his country to develop atomic weapons before the Allies. In one scenario, Heisenberg simply makes an incorrect calculation of the critical mass needed. But this would be a strange error in so brilliant a physicist. So, Frayn suggests in the play's most controversial scenario, perhaps Heisenberg deliberately did not do the calculation because he knew—just like Möbius and Leo Lehrer—what it might lead to. This Heisenberg seems heroic in his self-sacrifice, his acceptance of the idea of limits to human knowledge.

From Conscious Theatricality to Performativity

Since Brecht's *Galileo*, there has been both a surge of new science playwriting and a distinct change in the way such plays are written and performed. This has to do with the idea of performativity and specifically with the integration of speech-act-driven dialogue into the contemporary science play.

Performativity emerged and evolved from its origins in speech-act theory as set forth by J. L. Austin in *How to Do Things with Words*, which "reminded us, especially, how vital to any successful act of communication are the interpretive conventions that govern it, and to which (to some degree intentionally) all parties to the act of communication must agree."[53] In this seminal book, Austin proposed different categories of "speech-acts," depending on what action they accomplish in their utterance. The classic example of a speech act is the traditional Christian marriage ceremony's exchange of vows. The marriage itself is performative in that it is achieved entirely through the language of the simple vows "I do." Austin's theories have interested a wide range of literary critics beyond the field of discourse analysis, some of whom have revisited the ambiguities and gaps in his original work in relation to literature. Both J. Hillis Miller and Annabel Patterson, for instance, have noted Austin's omission of literary speech acts. Patterson observes that "while Austin himself excluded literary communications from the category of speech acts, as being non-serious or as pretenses from which no action follows, others have subsequently felt that the line between 'real' and 'literary' communications cannot be so sharply drawn. It seems clear that the act of promising, for example, is logically indistinguishable from that of declaring one's intention, through the invocation of a Muse, to deliver an epic poem."[54] Contemporary science plays seem to be powerful examples of how fraught the boundary is between the "real" and the imaginary, promised, communication. The concreteness of the scientific enterprise, with its basis in ocular proof and known facts, vies with the nebulous, internal, and often subconscious motives and intentions of the scientist. The tension thus created makes for electrifying drama, despite the often heavy textual emphasis of science plays.

Let us take the play *Copenhagen* as an example. At first glance, a play like *Copenhagen* hardly seems concerned with performativity, let alone conventional theatrical methods. It seems to privilege textuality over theatricality in a script that, startlingly, lacks any stage directions except intradialogic speech acts (such as when Heisenberg indicates, "I crunch over the familiar gravel," or tells us that he is looking at Bohr and Margrethe). The lack of extradialogic stage directions seems to bear out the textual emphasis, as does the requirement that the actors mime opening doors and other actions that remove the play from the realistic performance sphere. Yet on stage, particularly under Michael Blakemore's direction and with Peter J. Davison's design and Mark Henderson's lighting, the play demonstrates its dependence on performance for the fullest exploration and conveyance of its central scientific metaphor. This is also the case with *After Darwin*; like *Copenhagen*, it is a heavily verbal play, yet it too relies on performance not only to demonstrate its scientific ideas but to enact them in such a way that the science is both performed for us and transformed into metaphor on the stage.

Just what is the notion of "performativity," and how is it relevant to science on stage? As Alisa Solomon, Jonathan Culler, and others have pointed out, the concept is frequently abused and misunderstood:

> In recent years, "performativity" has become the buzzword of postmodern academic theory, invoked to describe the constructedness of a range of "marginalized identities" or "subject positions": the particular phrase "gender performativity" has already become a hackneyed and much misused mantra of critical writing, its meaning becoming more diffuse even as the term acquires more buzz. As it has been seized by the critical industry, "performativity" has been uprooted from its origins in J. L. Austin's work on the relationship between utterance and action, and has been whisked into the realm of theatrical metaphor. There, the performative—"the reiterative and citational practice by which discourse produces the effects that it names"—has been translated, like Bottom in his ass's head, into a fancy synonym for performance.[55]

It is important to distinguish the performativity of contemporary science plays like *Copenhagen* from the simple demonstration of a scientific principle that can be found, for instance, in Brecht's *Galileo*, when, Galileo uses the stage to illustrate for Andrea his theory of the heliocentric universe in the play's opening scene, putting the iron washstand in the middle of the room to represent the sun and placing Andrea in a chair next to it to act as the earth. When Galileo picks up Andrea, chair and all, and moves him to the other side of the washstand, he neatly shows him (and the audience) how the Copernican theory works. A similar type of enactment, or literalization, occurs later, in the Street Scene (scene 9), as the Spinner spins around endlessly, mimicking the earth's movement; similarly, the Ballad Singer's wife dances around him, "illustrating the motion of the earth."[56] In both cases, science is performed, but the demonstration is purely didactic and not integral to the structure of the play; nor does it serve a larger thematic purpose.

By contrast, both *Copenhagen* and *After Darwin* are performative in the classic Austinian sense that they do the thing they talk about; they bring into being a material enactment of an abstract idea under discussion through a speech act. Put simply, they reflect "how to do things with words": in this case, words such as "evolution" and "the uncertainty principle." Much of the dialogue is in the present tense and describes actions, thoughts, or intentions in such a way as to enact or bring them into being. This also applies to the science in the play. Essentially, the dialogue of *Copenhagen* is one long speech act that performs the uncertainty principle in a way that only the liveness and immediacy of theater can bring about. Through their corresponding movement and speech acts the actors demonstrate that exact measurement of position and momentum is not possible, because the observer affects the act of measurement. The dialogue does not merely reflect the principle; it makes it happen, with the audience participating in that act of creation. This is why plays like *Oxygen* and *Proof* are not typical of the most successful science plays; the science (or in *Proof*'s case, the math) is superficially imposed on the play, so that the science and the theatricality bear no relation to each other and are not interdependent.

Speech-act theory is also relevant to science plays for another reason. J. Hillis Miller points out what he calls a "metalanguage" operating in Austin's seminal book: "Austin has a habit of commenting on what he is doing, to some degree from the outside, as though he were two persons, the one doing it and the other watching the first doing it."[57] Another term for this is Heisenberg's uncertainty principle, which figures centrally in several recent science plays. While an exhaustive examination of the link between speech-act theory and theater is not possible here, the aim is to suggest the ways in which the two interact and complement one another, perhaps as a point for further exploration.

Music and Light: Science Musicals, Operas, Farces, and Comedies

In recent years science plays have diversified into a wonderful array of styles and modes, and no discussion of the tradition of science playwriting would be complete without a mention of the significant number of new musicals and operas, as well as several farces and comedies, that deal with science. Most notable among these is *Fermat's Last Tango*, a musical that depicts Princeton professor Andrew Wiles, the English mathematician who proved Fermat's elusive theorem in 1992.[58] Like Faustus, he is pushing against the known boundaries of knowledge, in this case within his own field of mathematics. As the show opens, we see Wiles (called Professor Kean in the play)—refreshingly portrayed not as a stereotypical "geeky" mathematics professor, but endearingly genuine and "normal"— jubilant at having cracked Fermat's theorem, and vowing to his wife to start leading an ordinary life again. He is then devastated to learn from Fermat, who visits him in his study in a vision, that his intricate proof is flawed. Fermat leads him to the mathematical afterlife (the "Aftermath"),

where Kean engages in witty exchanges with great mathematicians of the
past: Pythagoras, Euclid, Gauss, and Newton. These eminent mathematicians come to Kean's aid in the face of Fermat's taunts that he will never
iron out the "big fat hole" in his proof. The musical is notable for its
successful integration of a surprising amount of "real" mathematics with
a charming and witty score.

Several other examples exist of this sort of science and theater event
that combines music with science in a lighter spirit than the straight
science play (see appendix). Music has its own mathematics, and it has
proved to be a powerful element in theater pieces that engage science and
medicine. This is true of the collaborative work of director Jean-François
Peyret and neuroscientists in France in which music features prominently
and enhances the experience of the audience's sensing of the science (see
conclusion), and also in *Viewing the Instruments*, a recent "medical and
musical collaboration" between musician-composer Philip Parr, gastroen-
terologist Dr. Peter Isaacs, and artist Jane Wildgoose. This piece uses
Marais's brief composition *Le Tableau de l'opération de la Taille* (1725) as
the starting point for musing on the experience of the patient undergoing
gallstone removal surgery, from the eighteenth century to the present.

The Anatomy Theater and the Science of Performance

Viewing the Instruments, which was actually set in an old anatomy theater,
raises the point that the interaction of science and theater is not confined
to the stage. There is a tradition of "performing" science outside the formal venue of the theater: one thinks especially of the anatomical theater
of the eighteenth and nineteenth centuries, with the literal staging of
surgical procedures and scientific experimentation creating "a highly theatrical event."[59] This goes back even further. "Although dissection of
human beings for anatomical study was revived in the late Middle Ages
after a centuries-long hiatus," writes Mary G. Winkler, "the modern science of anatomy was born in the Renaissance."[60] Then, as David Knight
notes, "in the anatomy theaters of Renaissance Europe, professors became
performers."[61] It has been shown that the methodology of the anatomical
theater was quite deliberately and self-consciously modeled on theatrical
performance: "experiments that always worked" were rehearsed before-
hand until they were sure to succeed, then "performed" as if the audience—which had paid admission for the show—were witnessing an actual
experiment that had not been seen or tested before.[62] What attracted

audiences to "the bloody sight of a body cut open and dismembered"?[63] Perhaps it was the chance to witness a theatricalized probing of "the essence of humanity. . . . Dissection, like the dramatic theater, offered visual spectacle with a verbal lesson: Listener, learn thyself!"[64]

Another trend in the interaction of science and the stage can be found in the "scientificization" of theater by various practitioners who have developed theories and approaches to performance that approximate scientific models. This is especially true of Brook, Stanislavsky, Brecht, Meyerhold, Grotowski, and Boal, all of whom have set down and codified their ideas and strategies about performance. Systems like Brecht's epic theater, Meyerhold's highly stylized biomechanics and Stanislavky's "Method" of naturalistic acting all to some extent borrow the aura of authority vested in science.[65] This also extends beyond acting to areas like lighting and scenery (Adolphe Appia and Harley Granville-Barker, for example). Brecht had always been fascinated by science, though, as we know, his relationship to it became increasingly fraught as he was developing his later work such as *Galileo* during the war and after. He presents his theories about Epic Theater in a way approximating a scientific enterprise or experiment. The focus is on the results he wants to achieve in the spectator; the actor and the stage are mere conduits. Meyerhold devised biomechanics in a similar way but with a much greater concentration on the performer, and specifically the performer's body. His understanding of the body's physiology draws heavily on science.

Even the way in which theater scholars are trained to research and publish their work derives from and emulates the scientific model of hypothesis, investigation, and evidence to substantiate one's ideas. In this regard, theater historians and practitioners are like artists in other media who have also attempted to turn their theories into a science. For example, a group of painters in the late nineteenth century closely related to the pointillists devised an elaborate system assigning desired emotional response to specific colors. The symbolists believed one could achieve a sort of synesthesia through the scientific analysis of the relationships between color, sound, and language. Bergson came up with a "science" of laughter. These developments all coincide with the rise of positivism (especially its direct impact on naturalism in the theater through Zola's transmission of the ideas of Comte, Sainte-Beuve, and Taine) and the authority of science during the late nineteenth and early twentieth centuries, and also with the development of new sciences such as psychology and the other social sciences. Havelock Ellis, Freud, and others began to approach sex from a more scientific angle, for instance. One might deduce, then, that all these nonscientists were seeking to legitimize controversial or marginalized subjects by lending them the authority of science. This, in turn, reflects the increasing prominence and status of science within culture.

Conclusion

Looking back over the tradition of science plays, we can trace specific currents in the emergence of the genre. Shaw, Brecht, Dürrenmatt, and other earlier playwrights take concrete problems and issues as the stuff of the scientific drama. Often, they are primarily concerned with the moral role of the doctor or scientist, the public responsibility as well as the personal pursuit of truth.[66] Many of the more recent plays, however, enlarge that pursuit by exploring other concepts, such as moral ambiguity and the unreliability of memory. In an age of remarkable advances in scientific knowledge, from mapping the human genome to elucidating brain mechanisms, contemporary playwrights seem paradoxically attracted to some of the scientific and mathematical principles that seem most mysterious and unstable, like uncertainty in quantum physics, chaos theory, and the neuroscientific basis of perception and memory. Plays about science tend to be quintessentially postmodern, in this respect as well as in their use of the stage. This is another aspect of science and theater that *Science on Stage* investigates.

There is no question, then, that science and theater have been interacting for aeons. But over the centuries, the *way* in which science has figured in plays has changed profoundly, and it is this aspect, as well as the sheer number of plays that deal with science, that is most interesting and significant. From Marlowe to the mid–twentieth century, we observe several tendencies that have helped to define the tradition. One is a "conscious theatricality" that keeps the audience aware of the connection between the form of the play and its content—between the theater and the science. This reaches a new level in the hands of Brecht, whose fusion of politically inflected scientific content and ideas with a theatrical form that helps to convey them serves as a milestone in the development of science playwriting. This in turn leads to a tendency to experiment with formal innovations that culminates in the newer science plays of the late twentieth and early twenty-first century.

Science plays have also traditionally shown an interest in depicting the conflicts inherent in scientific advancement, whether on the micro-

level of interpersonal relationships or on the macro-level of clashing institutions and systems of belief. Finally, a concern with "real" and substantive science over fantasy and over love interests is also a hallmark of
science plays. Whatever the scientific field, the best science playwrights
show a genuine familiarity with the ideas they are borrowing, a commitment to accuracy, and a willingness to risk departing from standard theatrical fare. Such flying in the face of received ideas about what audiences
will and will not tolerate in the theater has often proved unexpectedly
successful. The next chapter will explore the possible reasons for the popularity of "science plays."

2 Why Theater? The Appeal of Science Plays Now
·························

One of the main contributions of science plays is their collective challenge to the notion of realism, both in life and on stage. "What is reality?" a Doctor asks his or her Patient in Peter Brook and Marie-Hélène Estienne's "theatrical research," *The Man Who,* based on Oliver Sacks's neurological case study "The Man Who Mistook His Wife for a Hat." Science plays in general seem to be asking that question, probing "reality" and what science has told us about it over the recent decades, and doing so with provocative alternatives to straight realism, as if their questioning about reality should also be reflected in their use of form.

It is an interesting paradox of science plays that they are both "old" and "new" in their use of realism. Much of the dramatic core of science plays revolves around a conflict (or several conflicts), and since conflict traditionally has been the driving force of plays, science plays are traditional in this sense.

In Bertolt Brecht's *Galileo,* for example, we have the conflict between science and religion; the same is true of Timberlake Wertenbaker's *After Darwin* and Jerome Lawrence and Robert E. Lee's *Inherit the Wind.* In Brian Friel's *Molly Sweeney,* the three characters are in conflict with each other and with themselves over the operation on Molly's eyes and its consequences. In Margaret Edson's *Wit,* the conflict revolves around the medical team treating Vivian's cancer and her resistance to their inhumanity—and the conflict bleeds out to encompass the different approaches to life and death represented by the arts on one side and the sciences on the other. Physics plays—from the little-known documentary plays $E = mc^2$ and *Uranium 235* to the popular *In the Matter of J. Robert Oppenheimer* and *The Physicists* to the phenomenally successful *Copenhagen*—have ethical conflicts as well as personal ones, centering on the use of potentially lethal scientific discoveries. They pose dilemmas for the individual scientists, as well as on the much larger collective scale of all humanity, which makes their conflicts all the more powerful and dramatic.

Plays like these can be seen as classic and traditional in their use of conflict as central to the drama. Their action hinges on the unfolding of the conflict, and their suspense lies in how it will be resolved (or whether it will be at all). They also seem to lean toward the traditional notion of a central character as either hero or villain. Yet in their structure and their formal elements, science plays often depart radically from the expected realistic packaging that such dramatic conflicts usually come in: the most successful are not "well made" in the sense of having acts, scene divisions, intermissions, linear and causal plot development, and an intact fourth wall. Rather, they tend to opt for such forms as pastiche (*The Physicists*), or a hybrid of realism and direct address of the audience (*Wit*), or a juggling of two time periods each played realistically in its own right yet overlapping impossibly (nonrealistically) on the stage (*Arcadia, After Darwin, An Experiment with an Air-Pump, Oxygen*), or the seemingly antitheatrical, pared-down mode perfected by Frayn and Friel in *Copenhagen* and *Molly Sweeney*, respectively.

According to Nick Ruddick, "The science play [is] primarily realistic, though capable of incorporating fantastic elements."[1] Niels Bohr, his wife, Margrethe, and Werner Heisenberg in *Copenhagen* "have the ontological status of revenants, though they dress realistically and behave onstage as if they were still alive."[2] This ability to integrate realistic-looking and realistic-sounding characters with nonrealistic settings seems to characterize many contemporary science plays. Realism is not abandoned, but it is thrown into sharp relief and questioned. This happens in the works of Timberlake Wertenbaker as well as Frayn's. Asked once about the task of the playwright, Wertenbaker replied: "Very simply, to ask questions. . . . All a playwright can do is capture and phrase the questions as immediately as possible."[3] Modern art and literature tend to present situations or problems, pose questions, but not lay out clear-cut solutions or answers. To use a formulaic realism would seem contrived and false, as was already pointed out by the playwright Harold Pinter decades ago in his seminal essay on writing character:

> I suggest there can be no hard distinctions between what is real and what is unreal, nor between what is true and what is false. . . . A character on the stage who can present no convincing argument or information as to his past experience, his present behaviour or his aspirations, nor give a comprehensive analysis of his motives is as legitimate and as worthy of attention as one who, alarmingly, can do all these things. The more acute the experience the less articulate its expression.[4]

Pinter concludes that since characters' motives are often obscure, plays should not end neatly with a last-act resolution and an easy message. "To supply an explicit moral tag to an evolving and compulsive dramatic image seems to be facile, impertinent and dishonest. Where this takes place is not theater but a crossword puzzle. The audience holds the paper. The play fills in the blanks. Everyone's happy."[5]

Avoiding such a simplistic formula, many science plays employ a carefully crafted metatheatricality that constantly reminds the audience that they are in a theater, not eavesdropping on actors who are unaware of them. "Old conventions like the aside and the soliloquy take on renewed interest because they treat the audience as at one moment there and at the next moment not here."[6] Other distancing devices include using archetypal rather than individualized naming, as in *The Man Who*'s use of "Doctor" and "Patient"—this too is an old convention, from the allegorical figures of the medieval morality play (remnants of which can be found in the seven deadly sins in *Doctor Faustus*) through the archetypes of August Strindberg and the Expressionists through Brecht, whose characters represent historical forces more than "real" people. In the newer science plays, such devices appear in innovative guises, from the roller-skating representation of History in $E = mc^2$ to the Professor and the Actress in Terry Johnson's *Insignificance*. Some science plays, like John Barrow's *Infinities*, take metatheatricality to an extreme by physically displacing the audience: they employ the "promenade style" of theater in which spectators move about from one location to another, an effective way of keeping them aware of the workings of the stage and not succumbing to the illusion of reality.

Perhaps this experimentation with form is a reflection of the scientific method itself, revolving around experiments and the constant testing of hypotheses, questioning data, looking for greater awareness of how the world works. Certainly this would be consistent with the merging of form and content that distinguishes so many contemporary science plays. An interesting paradox, however, is that the reluctance to pander to an illusion of reality actually leads in many of these plays only to the illusion of a release from it—the playwright only appears to be ad-libbing or improvising, only appears to be out of control, while in fact the characters and situations are as tightly constructed and plotted as in any traditional play. Apparent mayhem and unpredictability are carefully scripted in.

One might say that the further science plays get from realism, the closer they get to real science; and the more compelling the science play, the more it tends to depart from straight realism.

Why Theater?

As we have seen, a high proportion of science plays portray scientists as antiheroes, perhaps even villains, and engage the dark side of the pursuit of knowledge. All this sounds a lot like Dr. Strangelove and his cinematic ilk. There have of course been films about science, but films that genuinely engage "real" or "hard" science are rare. Although some examples do exist, there are two reasons not to try to deal with them in this study. Film is a different medium, and I am focusing here on theatrical performance because the medium is as important as the subject matter—the form of the play, and the resources and conditions unique to live theater, are successfully and innovatively used to convey the science. In addition, when film deals with science, it tends to become science fiction. As one critic puts it, "Since the success of the *Star Wars* movies from 1977 on, sf [science fiction] has quickly been subsumed into sci-fi, a popular-cultural phenomenon dominated by Hollywood and characterized by the subordination of textual to visual values, of extrapolation to extravagance, of speculation to special effects."[7] One thinks of *Jurassic Park*, *Planet of the Apes*, and *A Beautiful Mind*. The film *Gattaca* is a rare exception in the degree to which the scientific ideas dominate the action and are intelligently conveyed.

Film has consistently failed to engage science as seriously as has the theater. Commentator Jim Burge identifies a "failure to respond to science" that seems to be specific to film:

> Since George Méliès, exactly a hundred years ago, made a film about a journey through space that ended on a moon with a face on it, the theoretical side of science has found it impossible to get into a movie plot. Science is either ignored or, when it gets in, provides what Hitchcock called a Mcguffin, something which is important for the story but has no need to be explained—the miracle drug, the secret formula. The awful truth is that a hundred years of scientific discovery have affected the world of popular narrative hardly at all.[8]

Very few films are able to deal with "real" science in the way that the theater has done; they turn the science into fantasy. What is it about the stage, then, that seems so much better suited to science?

Michael Frayn cites the interactive role of the audience in a dialogic process. In a seminar on *Copenhagen*, he explained that even by the end of the rehearsals for the London premiere the actors kept saying:

"Well, I still don't see why he [Heisenberg] came to Copenhagen." At that point they were also saying, as actors always do, at the end of a long rehearsal period, "Well, I don't think there's anything more we can do in the rehearsal room, we need to get in front of an audience now to find out about what's going on." And I said, "That is why." And I do think that the idea of human confrontation is absolutely of the essence, the whole of art, the whole of literature, the whole of storytelling, the whole possibility of language and communication. One can't communicate with oneself unless one communicates with others.[9]

C. P. Snow likewise uses the metaphor of dialogue to articulate his vision of how to unify the sciences and the arts and humanities. "Those in the two cultures can't talk to each other"; we must therefore hope for a "third culture" that would "be on speaking terms with the scientific one."[10] The theatrical experience is doubly dialogic; characters converse on stage, while in a larger sense the actors maintain an unspoken dialogue with the audience. The many recent science plays show how effective this multidimensional conversation can be, suggesting that the intersection of science and the stage may represent precisely the kind of "third culture" that Snow envisioned. This suggestion is especially appealing in light of the work of the influential French philosopher Michel Serres and his emphasis on thirds (rather than binary opposites), on a middle way that can unite supposedly antithetical elements.[11]

There is also the paradox of less being more. Nick Ruddick points out that "the stage imposes physical (and budgetary) limitations on the artist, and calls for simplicity where film allows visual, aural, and financial abundance. In addition, theatrical productions can accommodate far fewer viewers. This makes for an intimacy that is impossible in cinema, not just due to size but to liveness."[12] Perhaps it is precisely such limits that give theater its appeal. In a *New York Times* review of two productions of the science opera *Einstein's Dreams*, the critic Dennis Overbye suggests that, "if physics seems to be doing better on stage than on screen lately, maybe that is because plays require fewer resources. So they can afford to be experimental. Or maybe it's just that words, and thus ideas, are more important in theater than in the movies, a visual medium." Overbye wonders "whether the Janus faces of science and art might melt together more easily in the shadowy half-light of the stage, where a little greasepaint and our own conspiring imaginations help create the scene, than in the blinding information-rich literalness of celluloid."[13]

The emphasis on the visual, the tendency to excess, the lack of experimental or innovative forms: this all sounds like what Murray Smith calls a "casual condescension to film . . . suggesting that it is still regarded in some quarters as a glitzy but insubstantial upstart, a glamorous newcomer whose seductive appearance flatters to deceive."[14] The idea here is not to evaluate and judge one medium as superior to the other but to explore the possible reasons why theater tends to embrace science whereas film has curiously neglected it. Relying on limited resources means having to use our imaginations, which means becoming involved. As Frayn puts it, "The audience can be involved in a way that's not possible in cinema or television—and I think sometimes they do like to be asked to think very hard about what's happening on stage."[15] Witnessing a production, the audience "reciprocally affects, be it ever so slightly, the quality, tone, or atmosphere of the production in a way impossible in the movie theater."[16] The liveness creates a frisson of uncertainty about the outcome of any production, and a sense of shared experience between actors and audience—"a measure of unmediated authenticity that is precious in an otherwise highly mediated culture."[17]

One cannot ignore the possible socioeconomic factors involved in the intersection of science and theater as opposed to film. Theater audiences tend to reflect a higher level of education and affluence.[18] They tend to come from a wealthier demographic, as average theater ticket prices certainly indicate. They are also more likely to be familiar with the ideas and events depicted. One should ask, therefore, whether the presence of so much science in the theater is as much the result of economics and demographics as the high-minded and idealistic fulfillment of Snow's "third culture" vision. Is it simply an elite art form? Is that why artists have found theater more hospitable to science and therefore chosen that medium?

The answer is not an unqualified yes, because most science plays are by well-established authors who have written other plays that are not about science and have in fact made their name in drama through nonscientific subjects: Stoppard (*Rosencrantz and Guildenstern Are Dead*), Frayn (*Noises Off*), Friel (*Dancing at Lughnasa*), Wertenbaker (*Our Country's Good*), Churchill (*Top Girls, Cloud Nine*), indeed, most of the playwrights studied here. They are first experienced dramatists and second interested in exploring science. That they have chosen science is the key issue, not that they have chosen the theater as their medium, because that already *was* their medium. Their choice of scientific subject matter is connected to an interest in ideas generally, and in particular ideas relating to broad philosophical and epistemological questions that many have already explored in previous plays (Stoppard, for instance). The science proves attractive because it allows them a way of continuing this investigation into such questions as how we know what we know, how identity is constructed, what the ethical choice is in an immoral situation, what our responsibility is in our deployment of knowledge, how we can know ourselves and each other, and so on.

Science and Culture

As Glynne Wickham wrote in 1962—during the peak of the cold war and of the "two cultures" debate—we live in "a world where the individual has committed the control of his agricultural and industrial economy, of his communications, and even of his bodily and mental health, to scientists."[19] Because the theater "can provide a forum for the examination and discussion of the human condition, its relationship with its gods, and its interest in itself, collectively and individually," it might, he suggested, serve as a "safeguard to the tyranny of science," which has taken such "alarming . . . hold . . . on men's minds."[20] Four decades on, these sentiments still resonate. We have an uneasy relationship to science. Science and medicine are part of everyday life as never before, and are not just something for the experts; we are encouraged to go to the doctor armed with knowledge about our health gleaned from the Internet, books, or magazines. We read about cloning, genetically engineered food, cancer research, mental health drugs, the Human Genome Project, stem cell research, and multitudes of other scientific developments in the daily papers. Gillian Beer notes that "there is no secure divide between the domestic world and the laboratory. Both look like private spaces; both produce and are involved in communal consequences, world-wide transformations."[21] The media constantly foreground science stories, largely because modern science presents increasingly thorny ethical issues.

These issues can lead to a sometimes alarmist, sometimes wholly justified clamoring after greater illumination. Yet in the journey from scientific source to newspaper, magazine, or Internet report, serious omissions, compressions, and distortions can occur; not only the science gets misrepresented at times, but the very process of discovery. The fault lies not just with the journalists. Scientists are not always the best spokespeople for science, and Frayn maintains that scientists could better communicate the excitement of their enterprise: "I think that it has become fashionable to write up your results in as dull a way as possible."[22] The French neuroscientist Alain Prochiantz, who collaborates with theater director Jean-François Peyret, likewise finds the way scientists are trained to write up their results not only deadening but misleadingly opaque about the hesitation and uncertainties that characterize most scientific endeavor, but get erased in the process of publication. This is where the playwright steps in. By some alchemical magic, the fusion of skilled playwright with "dull" scientific material brings science to life on the stage.

Many science plays offer insights into the issues as well as the human faces behind them, and that seems to be part of their appeal. But there are many other aspects as well. As so many science plays demonstrate, the role of science within contemporary culture is a tense one, as new discoveries constantly create new ethical dilemmas. Now, as one commentator puts it, "Scientists are listening to their critics. They sense that the reputation of their profession has been damaged recently, by controversies over genetically manipulated organisms (GMOs), the combined measles, mumps and rubella vaccine (MMR), and the cloning of animals. The common element in all suggested remedies is that professional scientists should exert themselves to explain more clearly what they are doing, and why." There is an urgent need to "make science trusted again."[23] Scientists like Prochiantz corroborate this, saying that literature and science are farther apart than ever as the very notion of scientific "truth" is fraught. Furthermore, there is not enough transparency and openness about what scientists actually do on a daily basis. "We see nothing of the process of science," Prochiantz notes, "or the human beings inside the scientists; we just see the end results. Yet the process of science is highly individual."[24]

Perhaps science has become so "hot" in literature—both in drama and in novels—because of this need for illumination. Science is, paradoxically, at once ubiquitous yet still largely opaque and inaccessible. A degree of mediation is necessary in order to grasp the latest scientific, medical, or technological advances and their implications. Enter what Alan Lightman, author of *Einstein's Dreams*, calls the "longstanding love affair between scientists and artists," which continues to the present day.[25] Theater can both dramatize and demonstrate science in entertaining and immediate ways. "Art has always wrestled with emerging ideas. Science has always been a rich source for those ideas," Lightman notes. He suggests that "what science can offer art is that most subtle quality of all, the way that scientists think, the way they live in the world, or what one might call the mind of science. Science and art have different ways of thinking, and those differences, when explored and portrayed, can enlarge both activities." The different modes of thinking "represent different ways of being in the world. And, like travels to strange lands, they challenge our assumptions; they disturb us in creative ways."[26]

Shelagh Stephenson, whose *An Experiment with an Air-Pump* premiered the same year as *Copenhagen* and *After Darwin*, cites increasing secularization (at least in Britain) as the main reason behind the surge in science plays. "I think it is because there isn't religion anymore. . . . People now look to science for the answers."[27] Renowned theater director Peter Brook came to science for similar reasons. After his acclaimed staging of the Mahabharata, he was looking for a totally new direction in the theater; he "felt an intense need to move away from myths of the past, from historical subjects, from period costumes, from worlds of the imagination."[28] He looked to science for the simple reason that "today . . . we have a new mythology. Science explores the same eternal mysteries with a new symbolic language."[29] Brook was attracted to the theatrical possibilities inherent in Oliver Sacks's neurological case studies, leading eventually to a collaboration entitled *The Man Who*.

Writing about Dürrenmatt, Roger A. Crockett notes that "science has eliminated old superstitions, but . . . it has replaced old myths with a new, potentially deadly one: the myth of the infallibility of science."[30] Beyond the invention of a rich new mythic language and the substitution of new gods for old, there is the huge appeal of science itself. "It's such a suggestive area," says Michael Frayn, although "it has become extremely specialised, and it's hard for anyone who hasn't devoted a lifetime to the study of one particular science to know very much about it."[31] One would think such specialization would make understanding, let alone writing about, science as a nonspecialist rather daunting. But audiences and readers seem willing to undertake that task, and seem fired up by it. This is a fair indicator of the centrality—metaphysical as well as factual—of science to society. The interdisciplinary scholar and public intellectual Lisa Jardine thinks that "scientists occupy the place in our society once occupied by intellectual clergymen and armchair philosophers. . . . They offer the 'meaning-of-life' strand of thought in contemporary fiction."[32]

The general consensus seems to be that science has come to hold a profoundly important place in society, regardless of how we feel about its discoveries and controversies. In addition, it is a constant, self-generating source of new ideas, and that can be hugely appealing to artists and writers. Ian McEwan, a novelist who often incorporates science in his work, suggests that the built-in novelty of science makes it very attractive to writers looking for new material in their novels, because by contrast with science, "the domain of the novelist is human nature and endlessly exploring that," and human nature is not novel but timeless. Above all, science provides superb metaphors and images: poets and writers often "just raid science for metaphors."[33] The appeal of the handy, illustrative metaphor—whether the Möbius strip, the second law of thermodynamics, complementarity, uncertainty, chaos theory, or natural selection—can be seen in almost all the plays discussed in this book.

Yet many playwrights find the science secondary to the human dimension: people and their histories. Ken McMullen suggests that scientists and artists are "concerned in their own way with the mystery and the reality of what makes us who we are." The idea is "to achieve an understanding of the universe of which we find ourselves a part."[34] Pointing to the example of Oliver Sacks, the neurologist whose vividly narrated case studies have inspired at least one film and two plays, Frayn argues that writers and scientists are united by the common enterprise of describing the world.[35] Both need to be keen observers of their surroundings and to find accurate yet imaginative ways of describing them. For example, Frayn's dramatic oeuvre is unified by his interest in exploring the problem of intentionality—"why people do or don't do things," he says in one interview.[36] This is the way society moves forward, no less: "It's not what has happened that makes the next thing happen, it's how people perceive what has happened."[37]

A common interest among many science playwrights is the concept of justice. Justice is "the major theme of all of Dürrenmatt's works"; in *The Physicists*, "justice is a question of what to do with the power that stems from scientific knowledge."[38] Timberlake Wertenbaker emphasizes how theater can be important in the ongoing "search for justice" that must continue at all costs, and science can be a fertile ground for this search, given its potential for both harm and good. "There have been a lot of mistakes in science too, but one of those mistakes can lead to a new discovery. You don't dismiss Ptolemy as a complete idiot because he thought the sun revolved around the earth. . . . Many of his other calculations were very good. I'm not a scientist or a political thinker, I'm a playwright."[39] Wertenbaker again insists that the playwright should not dictate or preach but "very simply . . . ask questions. I think you pose the questions in an interesting, dramatic way, and that's it. You formulate the question." In fact, justice can only be teased out if you *don't* offer a solution: "I think if you know the solution, you should be doing something else, you should be in politics or writing pamphlets or something. All a playwright can do is capture and phrase the questions as immediately as possible."[40] In this regard, as well as in theatricality and staging, contemporary playwrights who use science in their work are continuing to do what all great dramatic art has done: to frame difficult and important questions in an accessible and vivid manner, and allow the audience to draw its own conclusions as to the possible answers.

The History of Science

Not only science itself—whether physics, mathematics, biology, or chemistry—but the history of science is proving a vital source for new plays. In the search for interesting material, playwrights are finding the history of science to be a rich and largely unmined repository of characters, plots, and ideas. The history of science presents inherently dramatic material: "great feuds," high stakes, intense competition, and extremes of elation and disappointment.[41] It also bears the added attraction of being common knowledge, of belonging to us all, of being part of our cultural memory. Even if we might not understand the science that created the H-bomb, we can follow the events and people involved in the race for it. Even if the details of Darwin's research are obscure to most readers and audiences of a play about him, his life story and the implications of his discoveries and ideas make immediate and accessible drama. It is no surprise, then, that so many plays depict paradigm-shifting events and figures in science.[42]

History-of-science plays are often paradoxical: they use some very specific scientific question or situation, such as the meeting of Heisenberg and Bohr in 1941 or the catastrophic accident caused by Louis Slotin at Los Alamos, not for the science itself but as a springboard to exploring philosophical questions that touch all of us. The end of the cold war is particularly significant in the rise of the science play, and in determining which sciences it would favor. In a more general sense, science has entered quite deeply into the public awareness, as something with a direct bearing on our lives, and not remote, secretive, and mysterious. The rise of the field of the history of science parallels the rise of the science play. Such shifts and developments in institutional emphases also underline the impact of science, for example, producing whole new generations of students and scholars of the history of science, which in turn produce new audiences for science plays.

Science plays further tap into the vexed subject of society's relationship to science as articulated and shaped by art—a relationship that has changed drastically over the past century. To paint this history in very broad strokes: the modernist avant-garde in the late nineteenth and early twentieth centuries (Strindberg, German expressionism, surrealism, Dada) styles itself as a rebellion against science, insofar as it associates science with positivism, industrialization, and technology. It sees science as state- and capitalist-driven and as opposing artistic vision. For example, Karel Čapek's dystopian play about technology, *R.U.R* (the source of the word "robot"), gives us science fiction on stage with the scientist as "villain," as do many other plays about scientists during and directly after the modernist period. They imagine a world ruined by greedy, evil scientists out to destroy life and art. With the passing of the avant-garde—and also, one might argue, with the emergence of forms of art that actually embrace science and technology—the oppositional stance toward science has vanished.[43] At the same time, there is the poststructuralist influence that has helped to delineate what we call postmodernism. Michel Foucault and others helped bring about a thoroughgoing critique of the project of the Englightenment, and since a positivist notion of science is central to that project, plays that seem critical of science in terms of its assumptions of absolute truth might seem to reflect this anti-Englightenment stance. In short, and with the caveat that the relationship between science and the modernist avant-garde is far more complex than this summary can possibly convey: whereas modernism tended to treat science with suspicion bordering on paranoia (Čapek, Huxley, Treadwell), more recently artists have taken a more open and curious stance. The avant-garde of modernism often displayed a hostile contempt for science, a feeling that it was at odds with both humanity and art; you were not truly avant-garde if you did not take that attitude. Postmodern artists, however, have more and more embraced science and technology as not only compatible with artistic aims but effective in helping to communicate them.[44] Many of the theater companies or groups that were founded in the late 1960s or after and are widely considered to be at the forefront of postmodern performance, from the Wooster Group to the Ontological-Hysteric Theater to Theater de Complicite to Tf2, are now prominently represented on the Internet and largely through their own Web sites. Their questioning, provocative stance toward mainstream values and dominant cultural norms may seem to be at odds with what may, in some ways, seem to be a capitulation to the very ideologies that they interrogate. This raises interesting questions regarding their relationship to this kind of technology—and whether their avant-garde status is compromised by being subsumed under the global marketing mechanism of the World Wide Web.

Ethics

Greater public awareness of and inclusion in scientific debates and developments perhaps results in a demystifying process. This development comes directly out of a postlapsarian, post–World War II consciousness, a sense of culpability as well as suspicion: we know what harm science can do to humanity, and as human beings we are all guilty if destruction rather than progress and benefit is the outcome. Our curiosity about new developments in science is now always tinged with anxiety; it is not just a matter of letting the scientists get on with it, since we know what "it" can lead to. This accounts for the great deal of public fretting about ethical dimensions of science, whether it be cloning, genetic manipulation, or controversial new cancer treatments.

The heightened role of ethics in discussions of science and medicine deeply connects these fields to the theater, since at some level most dramas have a concern over moral problems, and none more so than science plays with their inevitable ethical tangles. One critic has argued that the main purpose of science plays is to reestablish a sense of morality in the face of postmodernist relativism: "These plays are concerned to show that there remains a right and wrong way for people—including scientists—to act, even though the structure of physical reality, not to mention mathematical reality, has been revealed to be fundamentally indeterminate."[45] On the other hand, others argue that science plays are not inherently conservative as this critic claims, that they indeed reflect and celebrate the destabilization of truth.

However, the claim that science plays represent a search for some new moral vision—a "quantum ethics" that is consistent with moral complexities of modern life[46]—does not stand up against the evidence of the texts themselves. While many science plays do present ethical quandaries that are peculiar to contemporary science, such as cloning and genetic manipulation (motifs in plays like *An Experiment with an Air-Pump*, *A Number*, and *Safe Delivery*), their authors seem to be suggesting not some futuristic new morality but a recognition of accepted moral values. They seem to be saying that no matter how complex and thorny the moral issues are in the new scientific debates, "right" and "wrong" still apply and still mean the same thing; they may just be harder to discern. This is not so much a conservative turn as a sign of impatience with the excesses of postmodernist relativism.[47] It might also help to explain the renewed interest in documentary theater and in docudramas both on stage and on screen as discussed in chapter 3.

Powerful Personalities and Different Desires

As the previous chapter on the tradition of science plays illustrates, science plays are full of powerful characters. Strong personalities make great stage figures, whether they are real historical people or entirely made up. Christopher Marlowe's overreaching Doctor Faustus, Ben Jonson's cunning alchemists, Bernard Shaw's devious doctors—these characters are fictitious, although perhaps based on actual models. Increasingly, however, modern playwrights have seen that real scientists present ideal dramatic subjects. Galileo, Darwin, and famous modern physicists like Einstein, Bohr, and Feynman have been favorite subjects for modern plays.[48] Brecht's *Galileo* led the way in being one of the first science plays to feature a "real" scientist, and as we have seen, his depiction of Galileo continues to rile historians of science. Since Brecht, a wonderfully diverse range of scientists like Bohr, Werner Heisenberg, Ernest Rutherford, Marie Curie, Lise Meitner, Ralph Alpher, Sigmund Freud, Carl Jung, Rosalind Franklin, Thomas Huxley, Tycho Brahe, Johann Kepler, Stephen Hawking, and P.A.M. Dirac has peopled the stage.

Such figures are irresistible partly because of their authenticity, and because of the liberating quality of their often unconventional behavior. Quirky, idiosyncratic behavior can be deeply appealing and cathartic for ordinary mortals in the theater to watch. Unlike the rest of us, these characters are geniuses; and their genius at once produces and excuses their unconventional behavior. As Oppenheimer says in scene 5 of Kipphardt's docudrama about him: "People with first-class ideas don't pursue a course quite as straight as security officers fondly imagine. You cannot produce an atomic bomb with irreproachable, that is, conformist ideas. Yes-men are convenient, but ineffectual."[49] The same is true of artists. Timberlake Wertenbaker puts it this way: "Artists defy authority. You can't be an artist if you won't defy authority, and a lot of people who defy authority are in prison.... If you accept authority and orthodoxy, you cannot be creative. The whole point of being an artist is to look beyond the received ideas and to question them. I guess the more you question, the more you risk getting into trouble."[50]

Along with the attraction of dealing with real-life, exceptionally gifted, and therefore unusual characters, the realm of science provides refreshingly different motivations from the standard dramatic fare. Allen E. Hye notes in his study of the scientist in modern drama that science plays are united by their propensity to "view man [sic] as an intellectual creature. . . . it is the portrayal of humans as inquiring, intellectual beings— what can be called 'scientific man'—that gives these works their distinctive shape."[51] Their passion for ideas distinguishes the characters in science plays, signaling a revival of the theater of ideas of the nineteenth century and a receptivity for intellectually challenging drama on the part of contemporary audiences.

Dramatic characters are usually driven by some tangible, concrete desire—they want love, revenge, money, sex—that can be clearly identified by the audience as a thing to be gotten. The characters in science plays are motivated by desires, too, but usually of a more abstract order—a desire to know more, to be the first to know it, and to solve particular intellectual problems. "The hippest thing, any scientist will tell you, is the thrill of the intellectual chase and the discovery itself, the moment you alone know something new about the genome or the 11th dimension."[52] The desire is no less passionate for being centered on something more abstract than money or sex, though of course this is not to say that the scientists in science plays are above such passions—simply that they (and the plot) tend to be driven in the main by these other desires. In fact, the usual dramatic emphasis on relationships, particularly love interests, hardly figures at all in science plays. Love is generally unimportant.

Problems in conventional dramas are seen in negative terms; problems in science plays are things to be relished. This is true to the nature of scientific pursuit. As Peter Langdal, the Danish director of *Copenhagen*, has said, "I noticed that physicists, when they work together—if they've had a hard day at the office or wherever they work—love to go home smiling to each other, saying, 'Yes, we have a problem!' But when actors leave rehearsal, they think, 'Oh, I've got a problem.' "[53] The difference between the scientist characters in science plays and the characters in conventional dramas perhaps lies in this approach toward problems; for the former, problems are what keep them alive, not hurdles to be struggled over.

In *Copenhagen*, for example, the very words the characters use to describe their reactions to each other's scientific theories and discoveries convey this passion, as in the increasingly heated exchange (quoted in full here) about the merits of Schrödinger's wave mechanics:

HEISENBERG: . . . We're going backwards to classical physics! And when
   I'm a little cautious about accepting it . . .
BOHR: A little cautious? Not to criticise, but . . .
MARGRETHE: . . . You described it as repulsive!
HEISENBERG: I said the physical implications were repulsive. Schrödinger
   said my mathematics were repulsive.
BOHR: I seem to recall you used the word . . . well, I won't repeat it in
   mixed company.
HEISENBERG: In private. But by that time people had gone crazy.
MARGRETHE: They thought you were simply jealous.
HEISENBERG: Someone even suggested some bizarre kind of intellectual
   snobbery. You got extremely excited.
BOHR: On your behalf.
HEISENBERG: You invited Schrödinger here . . .
BOHR: To have a calm debate about our differences.
HEISENBERG: And you fell on him like a madman. You meet him at the
   station—of course—and you pitch into him before he's even got his
   bags off the train. Then you go on at him from the first thing in the
   morning until last thing at night.
BOHR: *I* go on? *He* goes on!
HEISENBERG: Because you won't make the least concession!
BOHR: Nor will he!
HEISENBERG: You made him ill! He had to retire to bed to get away from
   you!
BOHR: He had a slight feverish cold.
HEISENBERG: Margrethe had to nurse him!
MARGRETHE: I dosed him with tea and cake to keep his strength up.
HEISENBERG: Yes, while you pursued him even into the sickroom! Sat on
   his bed and hammered away at him!
BOHR: Perfectly politely.
HEISENBERG: You were the Pope and the Holy Office and the Inquisition
   all rolled into one![54]

The language of this passage ("repulsive," "madman") and the actions it describes show how intensely the characters care about ideas: people falling ill through sheer emotional strain, pursuing one another, "going at" each other, acting like "the Inquisition." Elsewhere in the play Heisenberg refers to Bohr's intellectual intensity: "You don't realise how aggressive you are. Prowling up and down the room as if you're going to eat someone."[55]

Similarly, in Tom Stoppard's *Arcadia* the "investigator" characters (Hannah, Bernard, Valentine, Thomasina) are driven by sheer love of knowledge and ideas. Those like the literary critic Bernard whose pursuit of knowledge is deeply colored by a desire for fame and money clearly occupy a lower moral ground. The passion for ideas is as visceral in this play as in *Copenhagen*. Thomasina is seriously distraught, even anguished, at the notion of the library of Alexandria burning down: "Oh, Septimus!—can you bear it? All the lost plays of the Athenians! Two hundred at least by Aeschylus, Sophocles, Euripides—thousands of poems—Aristotle's own library brought to Egypt by the noodle's [Cleopatra's] ancestors! How can we sleep for grief?"[56] The words "bear it" and "grief" underscore how deeply she feels this loss. For all his nonchalance, her tutor Septimus shares this passion and indeed channels his grief over Thomasina's death in the fire (which of course echoes the burning of Alexandria) into a quest to solve the mathematical problem she had embarked on but not finished—a passion that turns him into a hermit in single-minded pursuit of the solution to this one Herculean intellectual problem.

The passion for ideas is no less fervent in the modern scenes. Ironically, however, the emphasis is on how little is known despite the passage of time. Valentine declares how his work on algorithms "makes me so happy. To be at the beginning again, knowing almost nothing. . . . It's the best possible time to be alive, when almost everything you thought you knew is wrong."[57] This last statement follows Valentine's long speech explaining in detail the notion of chaos and predictability, delivered to Hannah, one of the most idea-driven characters in the play. Through this paradox of having the modern characters feel they know less than their predecessors, Stoppard demonstrates the notion of "heat-death" and entropy that are at the center of the play. The trajectory of human knowledge seems to be one of loss, from the optimistic Enlightenment scenes to the postmodern characters' flailing in the dark attempting to recover lost knowledge. Stoppard thus deliberately reverses the accepted wisdom about the steady march of progress.

What is interesting is how consistently in this play Stoppard shows such intellectual exchanges between males and females to be far more erotic, far more capable of igniting romantic passion, than physical attraction:

BERNARD: Why don't you come?
HANNAH: Where?
BERNARD: With me.
HANNAH: To London? What for?
BERNARD: What for.
HANNAH: Oh, your lecture.
BERNARD: No, no, bugger that. Sex.
HANNAH: Oh . . . No. Thanks.[58]

This completely unsexy conversation about sex serves to underscore where the real passion is in the play: between the researcher and his or her discovery. Physical attraction may be "the one that Newton left out," to paraphrase Valentine's remark,[59] and it certainly is important, but in a negative sense in that it can throw a wrench into an otherwise ordered unfolding of events. It is mainly of statistical interest. *Mental* attraction—whether between people or ideas—is what really matters, and where the sparks ignite. Bernard only proposes sex to Hannah because he has been turned on by her intellectual sharpness, which is a kind of narcissistic desire, since it mirrors what he loves most about himself. Septimus is beginning to be seduced by his pupil's passion for ideas as much as by her physical charms; Valentine and Hannah's dialogue is sexually charged because of the passion each displays about his or her research. But nowhere in the play is the passion for ideas expressed more directly than when Hannah declares, in a statement that might serve as the play's motto, "It's *all* trivial—your grouse, my hermit, Bernard's Byron. Comparing what we're looking for misses the point. It's wanting to know that makes us matter. Otherwise we're going out the way we came in."[60] Even Thomasina's desire to have Septimus teach her how to waltz—the closest the play comes to a conventional love scene—is first and foremost an expression of a desire for knowledge (to learn how to dance, or to kiss, is first an intellectual attraction and then a physical desire), only developing into romance once the intellectual affinity has sparked an erotic flame.

Conclusion

I have tried to suggest some of the contributing factors to the surge in new plays about science; undoubtedly there are many more. Most of these, it should be noted, are positive in nature, but we might also address some of the factors in the success of science plays that have a more negative aspect. The love of science for science's sake can also take darker undertones in science plays. This Faustian aspect can lead to such atrocities as eugenics and the proliferation of atomic weapons, leading in turn to the constant threat of annihilation, and it is a very prominent theme linking almost all the science plays I will be discussing and common to all kinds of science playwriting, from the Renaissance to the present. It is introduced in *Doctor Faustus*, whose very thirst for utmost knowledge is his soul's darkness. It suffuses Brecht's final version of *Galileo*, the one now most widely read and performed, wherein the scientist is castigated as an antihero (even while he questions the very idea of heroism). It is at the center of *The Physicists*, whose cold war context lends an even stronger sense of darkness to the play and its preoccupation with the inability to "unthink" a thought once it is thought—like so many science plays, this one asks us to consider the moral responsibility of the scientist with regard to the deployment of his or her ideas. This darkness lurks around the edges of even those science plays that are most optimistic and upbeat about scientific progress and discovery. It might be called, in a term that forms one of the motifs of *Copenhagen*, "Elsinore"—"the darkness inside the human soul."[61] Many science plays seem to be exploring the duality that science presents, the way in which it can be both "Elsinore" and Eden, so to speak. Perhaps part of the appeal of science plays is their ability to dramatize and stage this dialectical aspect of human knowledge and scientific progress.

A final, compelling reason for the appeal of science on stage may be what Gillian Beer identifies in Tony Harrison's science-based theater piece *Square Rounds* (about chemist Fritz Haber): the possibility of a "healing transformation."[62] Transformation, Beer notes, is "the most familiar and fascinating life-experience that we all share."[63] *Square Rounds* demonstrates this idea in breathtaking scope, in myriad theatrical moments involving magical transformations before our eyes and in the dialogue and lyrics constantly engaging this theme. In Harrison's play, the figure of Fritz Haber is the instrument of both good and evil, a gifted chemist whose development of fertilizer will "make infertile fields and wastelands flower," thereby battling worldwide hunger, but who will also help to develop the gas that killed so many troops in World War I.[64] Beer notes the appropriateness of the theater for exploring these ideas, since the liveness and immediacy of the stage enact a kind of transformation as audiences and actors engage one another. As Beer describes it, "In theater everything becomes possible."[65] As I hope the following discussions will show, the science plays addressed in this book bear this out as they demonstrate the transformative power of theater in dialogue with science.

Nowhere is this more profoundly addressed than in plays that engage physics. I will therefore devote the next two chapters to exploring "physics plays"—first looking at a selection of physics plays that are more or less from the middle of the twentieth century, and thus "early" examples of the genre, and then examining in some depth what is perhaps the most successful, widely recognized, hotly debated, and significant physics play of all: *Copenhagen*.

3
"Living Newspapers" and Other Plays about Physics and Physicists
●●●●●●●●●●●●●●●●●●●●●●●●

> JÖRG-LUKAS: I want to be a physicist, Papi.
> *Möbius stares at his youngest in horror.*
> MÖBIUS: A physicist? . . . You mustn't, Jörg-Lukas. Not
> under any circumstances. You get that idea right
> out of your head.[1]

Physics in general, and quantum mechanics in particular, has provided rich metaphoric material for writers, and not just of plays. Novelists from Ian McEwan to Martin Amis to John Kessel have drawn inspiration from this field. Nick Ruddick has argued for the way in which it can suggest a new "quantum ethics" that he argues is appearing now in contemporary science plays.[2] For David E. R. George, the relatively new field might even suggest a way of bridging the two cultures through its affinities with the theater. "It is not often that the theater has been taken seriously by scientists," writes George, "but few writers on quantum physics have been able to resist recourse to the theater for their similarly metaphoric descriptions of how the universe appears to them."[3]

Physics and theater have in fact been interacting for decades. *Copenhagen* may be for some the physics play par excellence (or indeed the masterpiece of all science plays), but many other plays have dealt with physics, some of which will be examined in this chapter. Physics plays make up the bulk of science plays (a whole book has been devoted just to British and American plays about physics between 1945 and 1964),[4] and one can understand why. The subject is innately dramatic; it entails conflict and controversy, the threat of mass destruction, and the possibility of vast unifying theories unlocking "the secrets of nature" (parodied by Dürrenmatt, for example, in the idea of Möbius's "Principle of Universal Discovery" in *The Physicists*). Physicists are among the highest-profile scientists;

in a relatively short time, physics has yielded more famous names than most other scientific fields combined, and playwrights can be fairly confident of audiences recognizing the names of Newton, Galileo, Einstein, Oppenheimer, Heisenberg, Bohr, and Feynman. The fact that most of these names are also closely linked with political events shows the potent mixture of science and politics and the dramatic appeal of such a mixture for playwrights.

In addition, the pace of new discoveries in the field of physics is staggering. Few scientific disciplines have experienced such rapid developments and of such magnitude, let alone of such destructive potential, over a relatively brief period. Since most of these developments are fairly recent, contemporary audiences will have a strong cultural memory of them, and this must surely play a role in the proliferation of plays that have to do with physics and physicists. Above all, there is something dramatically appealing in the sheer scale of physics: the discrepancy between the vast implications, the towering explosions, the lethal potential of the science and the lone and tiny figure of the scientist against such a backdrop. This is nicely captured in *Copenhagen* when Frayn's Heisenberg describes the immensity of what the German physicists were attempting with the reactor down in the cave in Haigerloch. The story of course parallels the situation of the physicists at Los Alamos; the play *Louis Slotin Sonata* dramatizes this as the title character struggles with a mechanism housing a force beyond his control and that eventually, and terribly, overcomes him.[5] In both cases, mere people are grappling with materials and forces much greater than themselves. Such moments recall Faustus and his lonely, horrible demise. The drama is inherent in the situations; the playwright need only find ways of bringing it out.

Of the many physics plays, a few in particular utilize physics concepts in interesting ways and will be examined in this chapter. First I will look at some documentary physics plays: Hallie Flanagan Davis's $E = mc^2$, Ewan MacColl's *Uranium 235*, and Heinar Kipphardt's *In the Matter of J. Robert Oppenheimer*, all written within the first few decades after World War II. Then I will analyze Friedrich Dürrenmatt's *The Physicists* in some depth, as it is surely a highlight of the science play tradition, as well as a key physics play.[6] Finally, I will look at some more recent physics plays, including a number that engage the elusive theory of everything and, by implication, string theory. These various plays are all chosen for how they engage and incorporate science and for their contribution to the ongoing discourse about the role of science within culture.

It's Up to Us: Physics and the Federal Theater

In general, physics plays express concern about "essentially honest and well-meaning" scientists up against "the machinations of power politics."[7] This is especially true of plays in the aftermath of World War II. For example, Upton Sinclair's *A Giant's Strength* (1948) was "one of the first U.S. works of fiction explicitly to urge upon scientists their particular responsibility for persuading their government of the destructive power of atomic weapons."[8] At the same time, two other plays were doing the same thing on opposite sides of the Atlantic: Hallie Flanagan Davis's $E = mc^2$ and Ewan MacColl's *Uranium 235*. Both of these little-known but intriguing science plays balance the potential benefits of nuclear energy against a stark warning of its deadliness, using the full resources of the stage to do so. They make stimulating and informative theater, even more than fifty years later. In addition, their common approach, a form of documentary drama, makes them particularly interesting to investigate given the current debates about the playwright's use of history on stage, addressed in a separate chapter in this book.

As I shall discuss below, documentary theater's popularity declined in the 1970s, perhaps because the form seemed too didactic and too reliant on absolute truth and objective facts in a postmodern world where such notions were being challenged by poststructuralism.[9] However, as Janelle Reinelt argues, docudrama has taken on renewed relevance in a society that is information-driven and is looking for the security of verifiable facts. Since documentary itself is, in Attilio Favorini's words, "a central paradigm of twentieth-century art,"[10] the combination of theater as a medium and science as the subject matter is all the more potent for audiences of the scientific age.

$E = mc^2$ discusses and illustrates complex scientific ideas, using the "living newspaper" style that Flanagan Davis had made her trademark as the director of the Federal Theater Project (FTP) from 1934 to 1939. The FTP was, in Harold Clurman's words, an "extraordinary phenomenon."[11] In its brief but remarkable span (1935–39) it reached millions of people, operating in forty states and employing thousands of unemployed actors and other theater professionals who were hit by the Depression. It was "the first (so far, the only) nationwide government-sponsored theater in the United States," and the significance of its contribution to our theater "has to this day not been fully recognized."[12] Although Flanagan Davis wrote $E = mc^2$ after the demise of the Federal Theater Project, the play was modeled on the Living Newspapers that had made her and the FTP

famous. The FTP pioneered this form in America which Flanagan Davis
had seen on a trip to Russia and which she and Elmer Rice adapted to
their own purposes. They drew on actual news sources such as articles in
newspapers, journals, and magazines, transcripts of trials, and government
hearings. The idea was to present topical dramas that could be adapted
by future companies using contemporary sources, and the aim was social
activism and enlightenment about serious public issues, not "museum art."
Flanagan Davis has described the way the plays were written and their
objective:

> The staff of the Living Newspaper was set up like a large city daily, with
> editor-in-chief, managing editor, city editor, reporters and copyreaders, and
> they began, as Brooks Atkinson later remarked, "to shake the living day-
> lights out of a thousand books, reports, newspaper and magazine articles,"
> in order to evolve an authoritative dramatic treatment, at once historic
> and contemporary, of current problems. . . . The Living Newspaper from
> the first was concerned not with surface news, scandal, human interest
> stories, but rather with the conditions back of conditions.[13]

The teams of writers were encouraged to "experiment with new forms,"
while drawing on all sorts of old ones—Aristophanic comedy, Shakespear-
ean soliloquy, commedia dell'arte, and Asian pantomime.

The overall aim of the Living Newspapers became thoroughly Brechtian: to dramatize "the struggle of the average citizen to understand the
natural, social and economic forces around him, and to achieve through
these forces, a better life for more people."[14] The texts were meant to be
adaptable to future times and audiences by having total flexibility to
shed and add passages, update the source material, and so forth. The FTP
frequently engaged science and medicine in its Living Newspapers;
Spirochete focused on the problem of syphilis,[15] while *Medicine Show* (produced on Broadway in 1940) displayed the inner workings of hospitals for
the audience.

Didacticism, topicality, and textual impermanence have combined to
leave $E = mc^2$ and most Living Newspapers out of the theatrical canon,
but the play deserves consideration for its extraordinary marshaling of a
wide range of scientific material, political issues, staging styles, and ethical
concerns. The play is part allegory, part documentary. It features among
its extensive cast a character called Atom and a Professor who explains
the physics the audience needs to know. Much of the dialogue is taken
directly from contemporary news sources such as transcripts of hearings
of the Atomic Energy Commission. It also deftly incorporates intertextual

references to—indeed, often large segments of—other plays that audiences would have recognized, such as *Wings over Europe*. In putting all
this material together, Flanagan Davis was answering the call of scientists,
led by Albert Einstein, for a "national campaign to educate the public on
atomic energy."[16]

The play was produced at Smith College, where Flanagan Davis taught.
"[Although] written and produced by non-scientific people, the finished
script met with approval from scientists," notes the anonymous reviewer
for the Smith College newspaper. From its inception, the play had scientific authenticity. Scientists at Smith and elsewhere "helped assemble the
bibliography; they borrowed private letters and diaries from the Manhattan and other projects; they read the script in each of its many versions
in order to point out inaccuracies."[17] Although "certain bits of dramatic
necessity were slightly at variance with scientific fact, . . . even the scientists admitted that the value of dramatizing an idea so important as that
of atomic energy justified" theatrical license.[18]

The play also received attention not just for being scientifically accurate and dramatically viable but also for demonstrating that the Living
Newspaper was an "exciting theater form." Granted, it would mainly work
in nonprofit situations such as university theaters, and would therefore
serve primarily educational purposes. But this is a reflection not so much
of its content as of the market-driven forces of mainstream theater.
"Graphic journalism, or journalistic theater, whichever one prefers to call
it, was regarded critically as *avant garde* when it was effectively used ten
years ago in Federal Theater Productions," Clurman noted in 1949. Although the demise of the form was caused by its main exponent (the FTP)
being shut down, rather than because of its "intrinsic theatrical merit,"
the main problem was one of topicality; its content, which had to be
constantly updated, meant that "to revise one of these plays would be
almost more work than beginning anew."[19]

The idea that the text was deliberately unstable gives these plays an
interesting, and unexpected, postmodern twist. Documentary theater, for
a long time unfashionable, is enjoying a revival of sorts as its formal methodology and structural devices have suddenly become more relevant to
today's audiences. There is a new role for documentary drama "in the
postmodern millennium," writes Janelle Reinelt. "In theater and film
docu-drama had its heyday in Britain and the U.S. during the 1960s and
1970s. As time passed, it lost its edge as issues of authenticity and factu-
ality became increasingly problematized by new historicism and historiography, and within the arts, the ideological critique of realism and its re-

lated forms. Feminists, for another example, have engaged in a vigorous interrogation of the meanings of witnessing and testimony, developing critique as well as validation. Recently oral history and what is increasingly called 'memory studies' have problematized the power of documents as well as subjectivity."[20]

Reinelt is interested in instances of "public events becoming performance/performative in their force."[21] Her examples are of high-profile, controversial cases like the Stephen Lawrence murder trial in England (the unprovoked, racially motivated killing of a black teenager as he waited for a bus on a London street), or the prosecution of David Irving, the Holocaust revisionist. Analogous examples in the United States would be the Rodney King beating in Los Angeles, which set off racial riots, looting, and burning for days in that city, and the murder of Matthew Shepard in Wyoming by homophobic youths. In all these examples, the impact of the tragedy was seismically felt on whole communities, not just individuals. It is this communal aspect that can be most effectively addressed by docudramas. Reinelt suggests that the newfound popularity of the "seemingly antiquated form" of documentary theater relates to its "recourse to facts, to the materiality of events, to brute display of evidence."[22] What she says about documentary theater might equally apply to the surge of science playwriting as a whole:

> In a time of great epistemological skepticism, we find a reassertion of the link between knowledge and truth. A kind of millennial anxiety seems born of a despair of establishing any credible "case" for facts, and a fear of losing history to those who would attenuate its materiality through methodological skepticism. The appeal of the old-fashioned documentary may be that it meets a deep collective urge for the link between knowledge and truth.[23]

Reinelt sees this urge on the part of the audience for "the materiality of events, of the indisputable character of the facts," as having the potential for "an ethico-political revolt, as a demonstration of caring, engagement, and commitment . . . which had [sic] widely deteriorated in the contemporary West. "[24] She is careful to point out that this does not indicate some "new ethical consciousness emerging in contemporary culture."[25] Rather, we just seem to seek affirmation in facts where we used to distrust them.

So how factual is the play $E = mc^2$? Take as one example the treatment of the concept of the Atom. The reviewer of the Smith College production singles out the character of Atom for special praise. Atom is unstable and unpredictable, by turns charming and churlish. Davis and her team of writers are clearly trying to suggest the intrinsic scientific properties of the atom through such human behavior. However, in one of the few critical discussions of $E = mc^2$, Charles A. Carpenter argues that the "most problematic element" in Davis's play is precisely this allegorical character. The way she is depicted—as volatile and "hypomanic"—is inconsistent with the scientific reality of atoms, which are "not only invisible but static if not set in motion: sheer potential."[26] This is an important point, and one that speaks to the problem of accurately conveying scientific ideas and concepts in the theater. Since one of the main messages of the play is that nuclear energy can be used for both good and bad purposes, depending on how we apply it, "Atom must somehow reflect the 'dual personality' that the two-sided potential of atomic power implies. Her manner is supposed to be 'docile and meek' at times, 'hard and manic' at others, as if the nature of atomic energy itself differed when used benevolently or destructively."[27]

This does seem to be a serious flaw in a play whose aim is the accurate portrayal of recent scientific developments so that we can all make informed decisions about them. According to Carpenter, "Quite apart from the general lack of appeal of this frenetic, comic-book figure, its striking lack of congruity with the concept it represents would disturb discriminating spectators and hamper the teaching function of the play."[28] This is a rare instance of form and content clashing in a play that otherwise gets its science right and in a tradition of plays about science that generally merge form and content extremely well. However, the very fact that Flanagan Davis has envisioned Atom as an allegorical figure who could convey the qualities of atomic particles through her behavior on stage suggests how close she comes to performing the science in the play, as in more recent science plays like *Copenhagen* and *Arcadia*. Atom's performance of her scientific properties may be inaccurate, but it is an important gesture nonetheless in the move toward staging science.

The legacy of epic theater in this play is quite apparent. Brechtian devices include direct address/breaking the fourth wall, overt didacticism, projections on screen, an episodic structure, an epic narrative span, and the use of allegorical figures or archetypes. The presence of a stage manager has the effect of dampening audience identification. Many of these devices will be recognizable to audiences familiar with Thornton Wilder's *Our Town* or *The Skin of Our Teeth*, or with Tennessee Williams's *The Glass Menagerie*, from roughly the same period; indeed, during the thirties and forties Brecht's influence was particularly strong in the American theater. In this case, the whole thrust of the play is to educate the audience about nuclear energy so that we can make informed decisions regarding its use, for example, when we vote in elections or sign up for electricity or do other practical, day-to-day things. The play seeks to make a link between the quotidian life of the people and the seemingly removed, esoteric science of nuclear energy.

The play's structure enacts its function: act 1 seems designed to impart information, while act 2 presents different scenarios of the uses and abuses of that information. The trajectory of the play is therefore to give the audience the scientific knowledge it needs to decide how nuclear energy should be utilized (or indeed if at all) in a post-Hiroshima world. In his docudrama *In the Matter of J. Robert Oppenheimer* (1964), Heinar Kipphardt has Oppenheimer state that "nuclear energy is not the atomic bomb. . . . It could produce abundance, for the first time. It's a matter of cheap energy. . . . plenty for all."[29] $E = mc^2$, written immediately following the bombings of Hiroshima and Nagasaki, also voices this hopeful aspect of nuclear energy, and it ends by leaving the fate of the earth in the audience's hands, pleading with us to choose the right path in our use of atomic energy. In an ironic twist, the play bears as its title a scientific formula, yet it clearly indicates that in choosing the fate of the earth there *is* no formula—we can only grope our way forward armed with as many facts as we can carry.

Various aspects of the play do stifle its chances of being performed in the commercial theater. The script calls for a cast of more than seventy-five characters and has thirty-five scenes. Many of the scene changes were supposed to be split-second ones. The material was then current but is now dated. These are some of the same characteristics of other docudramas of the time that attempted to educate the public about atomic energy. Arguably, it is the problem of all science plays, since they utilize material that stands a high chance of becoming outdated when there are advances in scientific knowledge, whereas the stuff of melodrama and domestic drama, such as dysfunctional families and their relationships, never goes out of style.

It is significant that Harold Clurman describes the Living Newspaper productions as "cinematic and journalistic" theater.[30] Surely this also describes the age we live in, when film and TV are the dominant media, and newspapers, magazines, and the journalistic sound bite our main sources of information. Douglas McDermott wrote in 1965 that "the documentary function once served by the Living Newspaper has now been filled by the swifter, more convenient medium of television."[31] Yet television has failed spectacularly to dramatize real science in fictional modes. Although it is a superb source of documentaries and nature programs, its potential to stage "hard" or real science in any way approaching the way theater has done has yet to be exploited. If Reinelt is right about the renewed relevance of docudrama to our own times, perhaps the Living Newspaper could thrive again, given a chance and the substantial updating of the material.

Indeed, Clurman's consideration of the FTP and its Living Newspapers concludes with an impassioned plea for theatrical rejuvenation by revisiting this form: "The theater [of the thirties] reflected what was going on in the world around it. . . . often verbose, hotheaded, loudmouthed, bumptious and possibly 'pretentious,' [it] did not produce communism in our midst; it produced a creative ferment that is still the best part of whatever we have in our present strangulated and impractical theater. We should look back to the groups, projects, collectives and 'unions' of the theatrical thirties, not simply with nostalgia, but with thoughts of emulation and renewal."[32] As McDermott points out, we should also note a great theatrical achievement of Living Newspapers: "the organic fusion of form and content."[33]

Another Warning to the World: *Uranium 235*

Ewan MacColl's collaborative, workshopped play *Uranium 235* represents a British version of the Living Newspaper. *Uranium 235* was actually written a year or so before $E = mc^2$ but was revised repeatedly over the course of its four years in repertory until 1952 and then revised again for publication in 1986. Though not as pure an example of documentary theater as $E = mc^2$, this work does at least loosely draw on historical sources. It is a pastiche of disparate scenes and theatrical forms unified by a Brechtian approach that uses didacticism and entertainment finely balanced through metatheater.

MacColl claims in his autobiography that hearing the news of Hiroshima and reading the Smythe Report (the Manhattan Project's report of August 1945, which made public for the first time the details of the development and potential of nuclear weapons) prompted him to write the play, sparked by suggestions from two colleagues.[34] Despite knowing "nothing about physics or, indeed, about science in general," MacColl wished, once he had been tutored in the subject himself, to write "a documentary on the history of atomic science from Democritus to Einstein."[35] As *Spirochete* did for syphilis, this play provides a history of nuclear energy from its humble beginnings in the concept of the atom through to the present day. *Spirochete* takes the audience on a highly entertaining romp through this dark chapter in the history of epidemiology and public health, and it uses all the familiar devices of epic theater and the FTP: allegorical figures, anachronistic dialogue, gallows humor, stark contrasts, and abrupt, rapid scene changes. Such techniques and devices are all present in *Uranium 235* as well. In epic structure, sheer theatricality, and physical demands on the actors, the two plays are very alike. It is certainly a remarkable coincidence that demonstrates the vital role of theater on both sides of the Atlantic in the postwar public discourse on nuclear energy. Theater did not simply reflect but actively helped to shape this public discourse.[36]

Nowhere in the published text does MacColl acknowledge any influence from the Federal Theater Project's trademark methodology of the Living Newspaper format. However, it is more than likely that such influence did occur, whether directly or indirectly, since Theater Workshop had "a documentary tradition going back to the thirties and directly influenced by the Living Newspapers."[37] Certainly the script bears this out in its formal as well as textual properties, particularly its direct injunction to the audience to make up its own mind about the use of nuclear energy. Most strikingly, *Uranium 235* attempts to make the form of the play in some way echo its content, so as to reinforce visually what is going on textually. It does what the Living Newspapers do: "utilize the total theatrical circumstances as a means of communicating to the audience solutions to its immediate and pressing problems."[38]

Like $E = mc^2$, the play uses personifications and allegories, for example, having Miss Mass ("a pretty girl wearing a spangled leotard") and Energy ("a muscular male figure") in similarly allegorical roles to Atom in $E = mc^2$.[39] As in $E = mc^2$, a stage manager—in this case, a scientist—guides the action, and this device is rendered even more metatheatrical in *Uranium 235* by having a subnarrator, the Puppet Master, take over this role in the second half of the play. The "Voice of the Living Newspaper" booming over the loudspeaker at times throughout the action of Living Newspaper plays is paralleled in *Uranium 235* by a "microphone voice" commenting on the action or facilitating transitions between episodes by setting the scene, much as Shakespeare's chorus does in the prologue to *Henry V*. The microphone voice also acts as an interpreter of the larger messages of the play. "This is only a play," the voice says at one point, "an attempt to discover the location of the audience's conscience."[40] Such distancing was all the rage during the 1930s and 1940s; Wilder and Williams, for example, often interject similar moments in their plays. *Uranium 235* also relies heavily on two cornerstones of Flanagan Davis's metatheatrical approach: audience plants (usually critical of and complaining about the play) and a great variety of visual and auditory forms. "Man in the audience" rudely interrupts the Scientist, for example, reminiscent of the hecklers in act 4 of Ibsen's *An Enemy of the People*. Actors are constantly commenting on the way the play is going, much as the actors do in Pirandello's *Six Characters in Search of an Author* or in Wilder's *The Skin of Our Teeth*, drawing attention to the play as a play.

Although performed on a bare stage, *Uranium 235* had "other things to stimulate the imagination, such as the amplified sound of machines, passing cars, railway trains [also used in $E = mc^2$], explosions, whispering voices, announcements of news items."[41] Both plays use large casts that have great demands placed on them in terms of the play's pace, with rapid costume and scene changes, and also their abilities as performers: they must sing and dance, speak in a range of accents and styles, and assume several parts in the course of the performance. In *Uranium 235*, as in $E = mc^2$, scenes are episodic rather than constituting a whole, unified plot or action. Transitions from one episode to the next occur quickly, and the contrasts in mode are deliberately abrupt, as when a scene characterized as "knockabout comedy," with Einstein, Planck, and Bohr trying to stage a ballet that would explain quantum theory, suddenly shifts into "'thirties gangster film" style, with Protons, Energy, Neutrons, Alfie Particle, and Lola the Smasher (Chadwick's neutron) talking like Al Capone.[42] Other science plays successfully utilize this form of episodic pastiche; Dürrenmatt hilariously parodies the Hollywood gangster film in *The Physicists* when suddenly, with no warning, the two spies masquerading as mad physicists pull guns on each other.[43] More recently, the innovative musical *Fermat's Last Tango* blends a range of musical styles such as patter songs, torchlight ballads, and operatic conventions. Such self-conscious mixing of theatrical modes is consistent with the tradition of science plays emphasizing theatricality.

Uranium 235 in fact opens where $E = mc^2$ left off: with an exhortation to the audience that the future is in our hands and a plea to choose the right path for nuclear energy. "We have conquered power and explored the innermost secrets of the origin of matter," says the Scientist. "There are no closed doors to us now. We can choose our own road and send fate scurrying before us like an idiot beggar. We have opened a door on the future and on that door is written Uranium 235."[44] The play comes full circle as it returns to this problem in the end, having demonstrated the history of the atom from its origins in Greek philosophy to its use in World War II, from Democritus to the alchemists to Bruno and Paracelsus, to Mendeleyev to the Curies to Einstein and beyond. After all this accumulated knowledge, we now have atomic weapons. Accused of "conspiring against the world, of betraying mankind to war and wretchedness, of using the brain to do the work of Death," the Scientist says: "The road that we have built across the wastes of ignorance is not a road which leads to Death except for fools who would throw themselves over the precipice. It is a good road which can lead to peace itself if only men will stop wearing blinkers on their eyes. It can lead to peace such as you have never known."[45]

The original ending concludes in exactly the same spirit as $E = mc^2$: placing the future of the earth in the audience's hands by showing them the different paths available in the use of nuclear energy and asking them to choose the right one. "There are two roads," declares Energy in the play's final moments. "Which way are you going?" is the last line of the play, and it is spoken in unison by All.[46] It is significant that, by the time the play was published thirty years later, MacColl had become highly dissatisfied with this ending and added to it an updated and far more cynical one demonstrating an extreme disillusionment with both politics and science. In this version, examples of nuclear reactor accidents are detailed, from Windscale in 1957 to the Three Mile Island disaster in 1979. "Each of those reactors is a weapon," a Woman says darkly.[47] And, just as in the last version of Brecht's Galileo, the figure of the scientist is castigated for selling out, for letting humanity down, indeed for compromising humanity for his own selfish ends. The woman says that "these dedicated scientists are as venal, as corrupt as . . ."[48]—it is left blank. The Devil? We are back in Doctor Faustus territory, where science is equated with corruption and evil. The Scientist symbolically removes his white coat, and the play ends with All plunged in darkness, whispering "forever" over and over again.

The story of the play's progress from its humble premiere to its success in the West End is remarkable. Uranium 235's first audiences were the patrons of Butlin's Holiday Camp at Filey in Yorkshire—"mums and dads and their children, as typical an audience of northern, working men and women as you could find."[49] Despite some of the company's concerns over the difficulty of the play's ethical implications, its scientific content, and its "foreign" style, the performances were a hit. To the gratification of MacColl and the company, the audiences "treated the play as they would have treated an exciting game of football. They cheered, groaned, shouted their approval, and when one of the actors tried to make a planned interruption from the auditorium they howled him down."[50] As MacColl records in his memoirs:

> It was a triumph and a complete vindication of everything we had said about the theater. A working-class audience could be won for a theater which concerned itself with the social and political problems of our time; furthermore, such an audience would accept any kind of experiment provided that what was being said continued to ring out loud and clear. In actual fact, what was regarded as wildly experimental by theater buffs and representatives of the theater establishment was accepted by our Butlin's audience as a perfectly sensible way of doing things. They were the radio and film generation. It is unlikely that more than a handful of them had ever been in a theater before, so why should they miss the French windows, the Tudor fireplace and the furnishings by Waring and Gillow?[51]

It is significant that this piece was once hailed as "wildly experimental." The peripatetic, homeless company took the show on a tour of one-night stands, and during rehearsals in Manchester they were discovered by Sam Wanamaker, the American actor who would go on to provide the financing for the Globe Theater in the 1990s. At this stage of his career, he had fled the United States, having "fallen foul of McCarthy's witch-hunters," and was himself on tour in an Odets play.[52] Excited by what he saw at the rehearsal for *Uranium 235*, he returned the next day, bringing the actor Michael Redgrave with him. "At the end of the rehearsal, they talked. Sam was wildly enthusiastic and said it was the best god-damned show he'd seen since the Federal Theater's production of Mark Blitzstein's *The Cradle Will Rock. . . .* [and] in less than two months we found ourselves playing under their sponsorship at the Embassy Theater in Swiss Cottage."[53] The three-week run was a huge success, and the company transferred to the Comedy Theater in the West End.

MacColl's play captures the fears and anxieties of many at the dawn of the nuclear age. It also reflects a need for facts, a feeling that the best preparation for this new world is objective knowledge. Like the other physics plays in this chapter, its educating zeal comes in a highly entertaining and theatrically successful package that performs the science in the play even while it is being discussed.

In the Matter of *In the Matter of . . .*

The last of our group of physics docudramas is Heinar Kipphardt's highly successful play *In the Matter of J. Robert Oppenheimer.* The play has become something of a cult classic among fans of both physics and documentary theater. Written in Germany during the frostiest years of the cold war, it was swiftly introduced to American audiences through its translation by Ruth Speiers and its much-publicized production at the Mark Taper Forum in Los Angeles. It is a remarkable science play in at least two aspects: its incorporation of real science and scientists (as *Copenhagen* would do much later), and its shrewd use of original source material.

The play is a documentary courtroom drama, much like *Inherit the Wind,* only adhering more strictly to its sources in determining the dialogue. The dialogue for the most part comes straight from the transcripts of the 1954 security clearance hearings of Dr. J. Robert Oppenheimer. Key players include State Prosecuter Robb, FBI investigator Rolander, and various scientists called to the witness stand, such as Evans, Bethe, and Rabi. The moral of the play is clear. As Roslynn Haynes points out in her discussion of it, "By judicious placement and selection of extracts from the hearing, Kipphardt has effectively investigated the investigating committee and transformed the original 'trial' of Oppenheimer into a trial of the U.S. government."[54]

There is hardly any actual science in Kipphardt's play; the closest he comes to anything scientific is the brief discussion of the effectiveness of the H-bomb in terms of what size area it would destroy. There is nothing else of a technical nature, and no physics terminology—in short, nothing that the layman would find difficult to follow or would need to have explained. In this regard the Kipphardt play is paradoxical: although profoundly concerned with the problems associated with science and the implications of research, it lacks the very science about which the characters are so concerned. The play is more about political, philosophical, and moral questions raised by doing science than about the science itself. For example, the issue of paradigm shifts comes up frequently. "The world is not ready for the new discoveries. It is out of joint," says Oppenheimer at the beginning of the play.[55] For Kipphardt, the hearings that form the drama of the piece clearly represent a paradigmatic shift in the public perception of the nature of scientific research and the role of science within society. "I cannot reconcile these interrogations with my idea of science," Evans (a chemist) confesses, "now that science has become so important. At any rate, I can see two kinds of development. The one is our increasing control over nature, our planet, other planets. The other is the state's increasing control over us, demanding our conformity. We develop instruments in order to pry into unknown solar systems, and the instruments will soon be used in electronic computers which reduce our friendships, our conversations, and thoughts to scientific data. To discover whether they are the right friendships, the right conversations, the right thoughts, which *conform*. But how can a thought be new, and at the same time conform?"[56]

These are some of the main issues the play introduces. But instead of repeating or synthesizing the many excellent analyses Kipphardt's play has received elsewhere,[57] I would like to focus on its similarities with other physics plays, particularly *Copenhagen*. Both plays are fundamentally concerned with understanding people's motives and recapturing the past. For instance, Rolander says, "I am trying to discover your motives, sir."[58] Robb (in a fictionalized monologue that ends scene 1) refers to Oppenheimer as "a sphinx" and says: "I have come to realize the inadequacy of being strictly confined to facts in our modern security investigations. How clumsy and unscientific is our procedure when, over and above the facts, we do not concern ourselves also with the thoughts, the feelings, the motives which underlie those facts, and make them the subject of our inquiries."[59] Frayn says much the same thing in *Copenhagen*, both in the play and in the postscripts, as when he notes its preoccupation with "the epistemology of intention." Later in Kipphardt's play Lansdale, who is recommending that Oppenheimer retain his security clearance, describes his pursuit of the physicist through conversations whose purpose was to probe him, "to find out what kind of a person he was, *what* he thought, and *how* he thought."[60]

This fundamental probing into one man's psyche is enlarged to a general question: "What kind of people are physicists?" When Evans asks Oppenheimer this, the latter responds: "You think they might be a bit crazy?" Evans says, "I have no idea, maybe eccentric; how do they differ from other people?" Oppenheimer replies: "I think that they simply don't have so many preconceived notions. They want to probe into things that don't work."[61] Such investigations seem to bring on the scientists' mistrust and suspicion.

A significant parallel with Frayn's *Copenhagen* occurs in Kipphardt's concern with recovering the past, as exemplified by the following episode:

> ROBB: In your letter, you give the gist of the conversation. Now I would
> like to ask you, Doctor, to tell us about the circumstances and, if possible,
> to give us a verbatim account of that conversation.
> OPPENHEIMER: I can only give you the substance, not the exact wording.
> It is one of those things I have often thought about . . . eleven years
> ago . . .
> ROBB: Very well, then.[62]

Oppenheimer proceeds to describe a visit by Chevalier and his wife to his
house in the evening—"I think he came for dinner, or for drinks"—but
he cannot recall exactly. So much hinges on Oppenheimer's recollection
of this visit, just as with the visit recollected in *Copenhagen*. Each play
highlights the unreliability and subjectivity of such attempts at recollection of events.

Later in the play, Lansdale lashes out at the "current hysteria over
Communism" that is responsible for the pursuit of Oppenheimer and so
many others, in which people are "looking at events which took place in
1941, 1942, and judging them in the light of their present feelings. But
human behavior varies in the changing context of time."[63] This is precisely Frayn's point in *Copenhagen*, and is borne out by the letters recently
revealed by the Bohr family (see chapter 4).

Oppenheimer and Heisenberg are counterparts; each directed the
atomic weapons programs of his respective country during World War II.
There is a distinct parallel between Oppenheimer's reluctance to participate in the development of the hydrogen bomb and Heisenberg's reluctance to help develop the German atomic bomb. In Kipphardt's play,
Robb quotes the letter by the Atomic Energy Commission that states that
Oppenheimer, from fall 1949 onward, "strongly opposed the development
of the hydrogen bomb; (1) on moral grounds, (2) by claiming that it was
not feasible, (3) by claiming that there were insufficient facilities and
scientific personnel to carry on the development, and (4) that it was politically desirable."[64] Many of these reasons also apply to the Heisenberg in
Frayn's play.

Kipphardt is eloquent about what it means to be a physicist. His choice
of dialogue indicates how alive he is to the problems facing physicists in
particular and scientists in general. Above all it is the tension between
pure research and its appropriation for military purposes that fascinates
him, just as it did Dürrenmatt at roughly the same time. The play also has
renewed relevance now because of its concern with the tension between

national security and individual freedom. "If we want to defend our freedom successfully, we must be prepared to forego some of our personal
liberty," states Radzi.[65]

The Physicists: Science at the Edge

Sacrificing one's personal liberty is a central issue in one of the most
famous of all science plays, *The Physicists*, by one of Kipphardt's contemporaries, the Swiss playwright Friedrich Dürrenmatt. This is perhaps the
best-known physics play before *Copenhagen*, and not surprisingly it exudes
a cold war sense of the menace of scientific advancement.

Three physicists, Einstein, Möbius, and Newton, are confined to an
asylum that is run by a hunchbacked spinster named Doctor von Zahnd.
They may or may not be mad; and they may not be physicists, as it turns
out, since at least some of them are spies whose "real" names are Kilton
and Eisler. One thing that is certain, however, is that each has murdered
his nurse with whom he has fallen in love. These terrible crimes serve as
analogies to the murderous potential of the science of physics, according
to Dürrenmatt. "I did it out of curiosity," says one of the physicists of his
science, "as a practical corollary to my theoretical investigations. Why
play the innocent? We have to face the consequences of our scientific
thinking. It was my duty to work out the effects that would be produced
by my Unitary Theory of Elementary Particles and by my discoveries in
the field of gravitation. The result is—devastating. New and inconceiv-
able forces would be unleashed, making possible a technical advance that
would transcend the wildest flights of fantasy if my findings were to fall
into the hands of mankind."[66] The play ends with a dark cold war vision
of evil triumphing over good and the three scientists addressing the audience in turn in a sobering series of curtain speeches.

The question of who is going to get this dangerous material drives the
action, along with a subplot in which a detective investigates the three
murders (and again the distinctions between science and murder collapse
as both come under the category of "criminal" behavior). Will the military
get the new findings first? The intelligence service? The politicians? Each
of the options seems unacceptable: Möbius cannot stand the idea of his
work being put to nefarious uses. But the inmate/patient Newton makes
the argument that the job of the scientist is to get on with his work and
leave the moral concerns to others. "It's nothing more nor less than a

question of the freedom of scientific knowledge. It doesn't matter who guarantees that freedom. . . . I know there's a lot of talk nowadays about physicists' moral responsibilities. We suddenly find ourselves confronted with our own fears and we have a fit of morality. This is nonsense. We have far-reaching, pioneering work to do and that's all that should concern us. Whether or not humanity has the wit to follow the new trails we are blazing is its own look-out, not ours."[67] The character named Einstein refutes this idea, saying that "we cannot escape our responsibilities. We are providing humanity with colossal sources of power. That gives us the right to impose conditions. If we are physicists, then we must also become power politicians. We must decide in whose favor we shall apply our knowledge."[68] These two opposing positions are summed up by Möbius: "All three of us have the same end in view, but our tactics differ. Our aim is the advancement of physics. You, Kilton, want to preserve the freedom of that science, and argue that is has no responsibility but to itself. On the other hand you, Eisler, see physics as responsible to the power politics of one particular country. What is the real position now?"[69]

After all this discussion of who should get Möbius's revolutionary manuscripts, he reveals that he has in fact burned them. He felt they would be too dangerous if they fell into the wrong hands, so he destroyed them, in the terms of thermodynamics, in a heat-death. He explains, "There are certain risks that one may not take: the destruction of humanity is one. We know what the world has done with the weapons it already possesses; we can imagine what it would do with those that my researches make possible, and it is these considerations that have governed my conduct."[70] Möbius says that his findings are so groundbreaking that "the consequences would have been the overthrow of all scientific knowledge and the breakdown of the economic structure of our society."[71] Rather like Faustus, "in the realm of knowledge we have reached the farthest frontiers of perception. . . . Our knowledge has become a frightening burden. Our researches are perilous, our discoveries are lethal. For us physicists there is nothing left but to surrender to reality. It has not kept up with us. It disintegrates on touching us. We have to take back our knowledge and I have taken it back."[72] But, as is true of entropy, this attempt at undoing knowledge, reversing the machine of progress, can only fail: a thought cannot be unthought, knowledge cannot be taken back, genius cannot "remain unrecognized."[73]

Möbius has concocted an elaborate, altruistic scheme to take back his knowledge. He feigned madness and renounced his family so as to sequester himself and his dangerous ideas in the safest place possible—an asylum. But his altruistic subterfuge backfires when it is revealed that none other than the head of the asylum, the twisted Doctor von Zahnd, has surreptitiously copied the manuscripts before Möbius burned them. Von Zahnd warns that attempting to hide discoveries and insights is a vain gesture: "He tried to keep secret what could not be kept secret. For what was revealed to him was no secret. Because it could be thought. Every-thing that can be thought is thought at some time or another. Now or in the future."[74]

Invoking the second law of thermodynamics as a theatrical metaphor is a strategy that later playwrights have found extremely effective; for example, both Tom Stoppard in *Arcadia* and Shelagh Stephenson in *An Experiment with an Air-Pump* use this notion in central ways. The second law of thermodynamics is explicitly introduced as a theme early in *Arcadia*:

> THOMASINA: When you stir your rice pudding, Septimus, the spoonful of jam spreads itself round making red trails like the picture of a meteor in my astronomical atlas. But if you stir backward, the jam will not come together again. Indeed, the pudding does not notice and continues to turn pink just as before. Do you think this is odd?
> SEPTIMUS: No.
> THOMASINA: Well, I do. You cannot stir things apart.
> SEPTIMUS: No more you can, time must needs run backward, and since it will not, we must stir our way onward mixing as we go, disorder out of disorder into disorder until pink is complete, unchanging and unchange-able, and we are done with it for ever. This is known as free will or self-determination.[75]

Thomasina's statement that "you cannot stir things apart" gives the lie to Möbius's doomed attempt to "unthink" his thoughts. We find this motif particularly in science plays that express concern about the morality of scientists and the responsibilities of the scientist toward the use of his or her work. In *An Experiment with an Air-Pump*, characters repeatedly assert that "once you've thought something, you can't unthink it, can you?" and "once you know something, you can't unknow it."[76] The echoes of Dürrenmatt reverberate in lines like "it's a continuum," in the many dis-cussions about whether science is morally neutral, and in Stephenson's creation of the scientist monster, Armstrong, a reincarnation of the twisted Doctor von Zahnd.

"Unknowing" knowledge was the original thematic and scientific cen-terpiece of the play. *The Physicists* was inspired by Dürrenmatt's reading of *Brighter Than a Thousand Suns* (1956), Robert Jungk's book about the development of atomic weapons that has since sparked such a range of interesting responses (especially with regard to Werner Heisenberg's appearance in it; see chapter 4). Dürrenmatt reviewed the book, "and its influence on him was significant," notes Roger Crockett. Dürrenmatt wrote: "The idea on which the atom bomb is based, the deep insight into the structure of matter, is a thought of the human race, represented as it were by a small elite group of researchers, and not to be appropriated by a nation. It is likewise impossible to keep secret that which is capable of being thought. Every thought process is repeatable."[77]

It is an indication of the overwhelming emphasis on texts rather than performance in the analysis of drama that no critic has commented on the significance of the Möbius strip as a performative in the play.[78] Yet surely this simple figure is not chosen randomly. The name "Möbius" conjures for most people not a man but an image: the Möbius strip, a surface that appears two-sided and two-edged but, when given a single twist and then conjoined, has only one side and one edge.

This image signifies the play's main ideas, and the truly remarkable thing is the way in which the play itself enacts or performs the symbol in turn. The Möbius strip metaphor is fully borne out later in *The Physicists*. It works because the play starts out by employing the usual, familiar dichotomies—tragedy/comedy, sanity/madness, goodness/evil, freedom/imprisonment—and then promptly subverts them, finally revealing that the distinctions between them are completely blurred. For example, the three physicists/spies toast their plan in dualistic terms:

NEWTON: Let us be mad, but wise.
EINSTEIN: Prisoners but free.
MÖBIUS: Physicists but innocent.[79]

But such distinctions collapse into a continuum when Doctor von Zahnd reveals her mad scheme to manipulate Möbius's findings and rule the world. The physicists' plan to maintain known dichotomies fails. Nothing in this play has two sides. The dualities on which we depend, indeed by which we define ourselves and our surroundings, vanish, leaving a very shaky foundation. Our dependence on dialectics to make sense of our universe thus crumbles. The physicists are even able to justify killing their nurses by calling them "sacrifices"; there seems to be no difference between murder and love.

The twist in the Möbius strip could be likened to the concept of "accident," which Dürrenmatt pinpoints as central to the play. *The Physicists* foregrounds the role of chance, or "accident," in determining our current situation as well as our fates. In his 21 Points to *The Physicists*, Dürrenmatt lays this out, especially in points 4 and 5, 7, 8, and 9: "Accident in a dramatic action consists in when and where who happens to meet whom"; "The more human beings proceed by plan the more effectively they may be hit by accident." In *Arcadia*, this concept of "accident" takes on the aspect of "noise" in the chaotic system, and is not only demonstrated in the action of the play but articulated by the characters. "The universe is deterministic all right, just like Newton said, I mean it's trying to be," says Chloë. "But the only thing going wrong is people fancying people who aren't supposed to be in that part of the plan." Valentine replies: "Ah. The attraction that Newton left out. All the way back to the apple in the garden."[80] Nothing can proceed in an orderly fashion no matter how hard we try, and both plays seem to characterize this as a particular attribute of contemporary life. Thus two scientific ideas, the second law of thermodynamics and the Möbius strip, receive performative treatment in highly theatrical ways in *The Physicists* and *Arcadia*, respectively, just as *Copenhagen* performs the concepts of Uncertainty and Complementarity. In the case of *Arcadia*, of course, the further implications of the second law give the play a gruesome twist, in that we know as we are watching that entropy is at work all the time, causing degradation rather than progress, the steady loss of information rather than the ongoing accumulation of it that Septimus optimistically envisions. This will be discussed in greater depth in chapter 6.

Strings and a Unified Theory of Everything

Now we come to the last group of physics plays—those dealing with the notion of an ultimate theory of everything, possibly realized in the recent concept of string theory. These plays are a long way from the docudramas of Davis, McColl, and Kipphardt and the tragicomedy of Dürrenmatt. They bring new meaning to the concept of science in theater and to the idea of performing science, since the science they engage deals in both the largest and the smallest scales of human understanding.

"Here, Faustus, try thy brains to gain a deity."[81] Could it be that what Faustus is seeking when he wants supreme knowledge—"these metaphysics of magicians"[82]—is none other than the elusive theory of everything? Perhaps the play is an early example of this prominent theme, one of the Holy Grails of science and a notion that has captured the imagination of dramatists. As we have seen, Dürrenmatt's *The Physicists* tragicomically foregrounds what might in contemporary terms be called a unified field theory or a theory of everything. Brenton's *The Genius* engages the idea in similar terms and, as we shall see, goes into more scientific detail. Both playwrights view it as the ultimate key to annihilation, threatening the very future of the earth.

Milder, more lighthearted "takes" on the unified field theory emerge in Terry Johnson's *Insignificance*, discussed later, and in recent works like *Humble Boy* by Charlotte Jones, *String Fever* by Jacquelyn Reingold, and *Calabi Yau* by Susanna Speier. Science figures in *Humble Boy* much as it does in David Auburn's *Proof*; there are just a few passages of scientific explanation scattered throughout the play, with little attempt to integrate the ideas formally. Apart from the scene (act 1, scene 3) in which Felix, a modern-day Hamlet, explains superstring theory to Mercy (whose response is "Oh! I like hearing all those funny words"), it is only superficially a "science play" and is concerned primarily with relationships in all their permutations—between parents and children, lovers, ex-lovers, and the living and the dead. This is precisely the same territory as in *Proof*.

The play *God and Stephen Hawking* (2000) also contains thematic references to a grand unification theory. It seems from the reviews that the playwright succeeds in making this a theatrical metaphor, in unifying form and content, by the device of having a single actor (Robert Hardy) play a wide range of characters with whom Hawking converses: God, Einstein, Newton, the pope, the queen. The stage has no limits; it has become the universe, the mind of God, *theatrum mundi*, a familiar trope in the history of science and the theater.

The idea of finding an overarching scientific theory that would neatly and simply explain how the universe works is understandably appealing, and has been eluding scientists for some time. Albert Einstein spent the last thirty years of his life searching in vain for "the so-called unified field theory" that would show that gravity and electromagnetism—two distinct forces—"are really manifestations of one grand underlying principle."[83]

As we will see later, Terry Johnson dramatized Einstein in the throes of his search for a unified field theory in the brilliant play *Insignificance*. Contemporary science writer Peter Atkins believes that "there will one day be a 'Theory of Everything': a theory, perhaps a set of equations, from which the properties of the whole Universe can be inferred."[84] Now some scientists think that they have solved the mystery of whether or not the universe can be described by one general, universally applicable law. String theory (or superstring theory) has developed over the last fifteen or so years to give us a kind of unified field theory that does what Einstein could not: it reconciles not only the different kinds of forces that exist in the universe but also the opposing laws of the vast (general relativity) and the minute (quantum mechanics). String theory "provides a single explanatory framework capable of encompassing all forces and all matter."[85]

As Johnson and other playwrights have observed, such a concept is not just scientifically attractive; it also has great dramatic potential. Rob Ritchie, the literary manager for the theater that first produced *Insignificance*, notes that "renewed anxiety about nuclear weapons triggered several plays in the early 80s that dusted down old-fashioned biographies of the great man [Einstein] or dispatched him to some blasted terrain to mumble apologies to crazed mutants."[86] *Insignificance* was one of the first Einstein plays to look beyond biography to the ideas themselves, not just to relativity but to the theory of everything as well. The image of Einstein seeking in vain for a unified theory of everything throughout his last decades does seem inherently dramatic, and the notion of an idea so universal is irresistible, appealing to some inborn Faustian need for ultimate knowledge about ourselves and our surroundings.

The appeal of string theory goes beyond individuals, however, for the quest for universality also provides a bridge between the two cultures. Like science, drama is often hailed for its universal themes and issues; indeed, that is often the litmus test applied to drama by critics, to see whether it will be a lasting, durable, "timeless" work. It should hardly be surprising, then, that string theory has appeared on stage in recent years, signaling its assimilation by dramatists interested not just in the latest science but in something that so neatly cuts across disciplines. Plays that allude to string theory and engage the idea of a theory of everything indicate both the popularity of "string theory" and the attraction in postmodern life of a theory that might comfortably explain everything, that would give us "the meaning of life." As Howard Brenton's play shows, however, this positive regard for such an all-encompassing theory is fairly recent.

Ominous Signs in the Snow

Brenton's play *The Genius* (1983) engages the notion of a "Unified Field Theory."[87] Strongly influenced by Brecht, *The Genius* features a modern Galileo in the form of a brash and brilliant young American physicist, Leo Lehrer, of whom Brenton writes: "Like Brecht's Galileo, Leo Lehrer cannot deal with the moral dilemmas his work forces him to confront. And, again like Brecht's character, this golden human being falls apart and becomes gross, in a 1980s manner. He is a dangerous man to know, arrogant, promiscuous, cruel and self-indulgent, a wrecker of the lives around him. I try to dramatize his reformation."[88] The other brilliant mathematician in the play, as in Tom Stoppard's *Arcadia*, is a female student with an extraordinary innate talent for numbers. Gilly Brown and Leo Lehrer "struggle with a dangerous idea—that nuclear science is a profoundly malign pursuit and that, for the first time in human history, we must deny ourselves a technological 'advance.' "[89]

Lehrer has evidently cracked the elusive unified field theory (what Möbius's theory of everything alludes to in *The Physicists*), but it has not made him a happy man. At the play's opening Leo has just left MIT for an anonymous, drab, redbrick UK university straight out of a David Lodge novel. The circumstances of this abrupt descent into hell are soon revealed: his work was picked up by the Pentagon, but he began to suspect their intentions for its uses and refused to cooperate. "My lack of human grace is brought on by a dose of the post-Einstein clap. Real guilt and dread. I had the new E equals MC squared but flushed it down the john. I feared it would burn the world."[90] Leo attempted what Dürrenmatt's Möbius also tried: to sequester himself and unthink his thoughts.

Leo says, "Galileo said one day, scientists will come forward overjoyed with a new discovery to be greeted with a universal cry of horror." The connection with *Doctor Faustus* seems clear: not just in the dilemma of wanting to test the limits of human knowledge but in the sense of doom or menace implicit in that desire to know everything. Will Leo's work "lead to a knowledge of nature which will endanger nature itself?" asks one university administrator. If the answer is yes, he says, "then we bury it. In the cellars."[91] It is no surprise that one of the play's intertextual motifs is Blake's "marriage of heaven and hell," frequently invoked by the characters. In addition, ghoulish theatrical effects include a blackout during which a glow-in-the-dark skeleton plays Bach's Musical Offering on violin, building to a tremendous explosion as a mysterious, unexplained symbol flashes across the stage. The glowing, violin-playing skeleton reappears at the end of the play. This recalls some of the effects in *Doctor Faustus* (particularly the B-text), with frequent use of fireworks, leaping devils, and so forth.

Although the idea of unified field theory is named already in the first act, Brenton waits until the two mathematicians are alone on stage to explain it. Lehrer has seen Gilly's equations in the snow (its purity at odds with the "dirty" implications of the mathematics), literally freaks out, and goes to great lengths to find out who she is and to arrange a meeting with her. In act 1, scene 3, he teaches her the implications of her equations by showing their relationship to physics, not just math. The audience, along with Gilly, is given a tutorial in the four forces of nature: "Gravity. The electrical force. The strong nuclear force. The weak nuclear force."[92] As he explains the link between the search for the force that unifies all these other forces and the international nuclear arms race, we watch Gilly undergo the same epiphany that struck Lehrer himself at MIT; "a gleam of darkness in the middle of all that light [the light of pure research]."[93] He tells her how "they wanted the work and they wanted me, for Uncle Sam, the free world, for weapons research, for—a—bomb. That's what it means, the tune you and I scrawled out with our ballpens. You describe how something lives and dear old human kind will use your words to kill it. *He shakes his head.* Oh boy, the consequence of describing life is death?"[94]

Overtones of several other plays occur at this point in the text when Lehrer reveals to Gilly what he did when this realization of destructive aims for his research sank in. "I gave up, Gilly, I closed down, I exiled me into my own head. If you are shit scared of the damage you can do, do nothing, eh?" He says he ran away to England, doubly exiled, but Gilly points out what was also a key theme in Dürrenmatt's *The Physicists* and continues to be prominent in plays like *Arcadia* and *An Experiment with an Air-Pump*: "But you can't not think. . . . A thought is a thought. You can't not have it. . . . You can't stop it, you can't stop nature."[95] As in *Arcadia*, the second law of thermodynamics serves as a handy metaphor for the irreversibility of human thought and progress. The possessors of those thoughts, however, are irreversibly damaged and haunted by their discoveries. In many of these plays we see the devastating consequences of "too much thought" on an individual scale, and the cases are tragic in the extreme. Leo Lehrer attempts something similarly heroic to Möbius by exiling himself to the Midlands, trying to render himself harmless by turning his back on the joint forces of the Pentagon and MIT. Both characters realize the futility of the individual against much more powerful institutional machines.

But what is most poignant and pointed in each case is not just these desperate attempts to unthink thoughts that have terrible implications. It is the human toll that such thought takes, as each character is changed forever by the force of his or her discoveries, giving new meaning to the term "postlapsarian." As Gilly says, "What do I do? Have breakfast? I don't think the egg on the plate will look the same. Ring my Mum? 'I know how they can make a new generation of weapons, Mum! Isn't that wonderful?' " Lehrer eventually replies: "Welcome to insomnia, sister."[96]

Far from Insignificant

Terry Johnson's hilarious and moving *Insignificance* shows a "Professor," who by clear implication is Einstein at age seventy, meeting an "Actress" who is obviously Marilyn Monroe, in the middle of the night in his hotel room. She is fleeing her tedious publicity obligations and her violent and abusive ballplayer husband (Joe DiMaggio), and she has sought out the Professor to demonstrate for him the theory of relativity ("Just the Specific Theory. The General Theory's a little too complex to go into here don't you think?).[97] She wants to be sure she understands it properly. The Actress proves to be more than a match for the Professor, fully up to the task of explaining his ideas. The ensuing demonstration is the play's pièce de résistance: the Actress does indeed explain the theory, in great detail and using all sorts of props, culminating in "two little trains, with track."[98]

The Baseball Player pursues the Actress to the Professor's hotel room; a fourth character, a bigoted Senator (Joseph McCarthy), has already tailed the Professor there. Thus the Professor and the Actress are in the same boat, each hotly pursued by a disagreeable and threatening male. The crass, foul-mouthed Senator is after Einstein to appear in court to testify before the House Un-American Activities Committee and name names. He recites dictionary definitions of words he has memorized, setting up a linguistic tension between the copiously used expletives, on the one hand, and the incongruously erudite vocabulary, on the other. This seems suggestive of many things the play is trying to convey: the tension between the inherent nature and the learned behavior, between what we are born with and what we acquire (nature versus nurture); the tension between the id and the superego (the Actress mentions Freud several times); the difference between knowledge and understanding. "Knowledge is nothing without understanding," says the Professor to the Actress when she tells him she memorized the whole general theory of relativity without understanding it.[99]

As we will see in chapter 5, the Senator has an analogous character in Shelagh Stephenson's *An Experiment with an Air-Pump* in the synonym-spouting Roget, and it is worth taking a closer look for a moment at what such wordmongers represent. Certainly in both plays they serve the simple function of providing some comic relief. But they also depict non-scientists embracing a scientific methodology in the organization of language, as if borrowing the authority of science. The Senator makes the audience think about meanings of words, their usage, and the issue of knowledge for knowledge's sake. His superficial use of language causes him to inhabit the other end of the intellectual spectrum from the Professor, who is seeking the most profound knowledge there is and for which there is not yet a language. Even the supposedly "dumb" Baseball Player in *Insignificance* is more knowledgeable than the Senator, and his anecdotes show his passion for the game, his understanding of it as if it were a science.

Against such concrete forms of knowledge, so to speak, Johnson pits the elusive theory of everything in the form of Einstein's insomniac quest, whose physical representation is the "pile of paper a foot tall" that is on stage from the play's opening. "It isn't just the culmination of a man's life work," the Actress tells the Senator, "it's a set of calculations that come close to describing the shape of Space/Time. He's almost unified the fields. If you'd just let him finish he'll have calculated how it all fits. How everything is."[100] The Professor keeps destroying, then adding to, and then destroying the work again. "I have finished my work four times," he says. "Each time I have destroyed the calculus and started over." The Actress is appalled: "But if you studied it, you'd know how it all fits. How the universe works. You'd understand everything!" Paradoxically, the Professor is undeterred in the fruitlessness of his endeavors: "I keep myself occupied. Mathematics is a splendid waste of time. I get to the end, I forgot the beginning. I go back."[101]

The Actress's passion for science and her affection for the Professor give the play a doubly delicious dramatic twist: not only do we have the improbable meeting of these two figures, but we have a characterization of Monroe that goes delightfully against type. As if to underscore this point, Johnson shows her at the very end of the play attempting to learn her lines for her current project, a film in which, as she says, "I take a pot-roast from the oven, I hear the doorbell, I run across the apartment removing my apron, I kiss the man, I disappear. No words."[102] The revelation of a deep and inquisitive intelligence beneath this "bimbo" image—an image reinforced by Johnson's incorporation of the famous skirt-around-the-ears moment as Monroe stands on the subway grate, a pose she has had to hold for hours before escaping to track down the Professor—makes these last lines so poignant, as the image once more takes over at the end and swallows up the real person we have glimpsed. Her phrase "No words" finishes the play, in stark contrast to the vibrant and chatty character we saw at the beginning.

Clearly, playwrights are divided as to the notion of a theory of everything. For some, like Brenton and Dürrenmatt, it represents the worst kind of Faustian bargain: ultimate knowledge at the expense of the whole of humanity, the future of the world. For others, like Johnson and Jones, the idea is less bleak. Their characters are not castigated for pursuing the idea, merely shown as rather ineffectual.

Spies and Lies: Uncertainty in Stoppard's *Hapgood*

While he is best known for plays like *Rosencrantz and Guildenstern Are Dead* and the "science play" *Arcadia*, which is analyzed at length in this book, Tom Stoppard is also the author of one of the first dramatic explorations of Heisenberg's uncertainty principle: the play *Hapgood*, written and performed a decade before *Copenhagen*. It is one of the most thoroughgoing theatricalizations of this scientific idea, but its relative failure on the stage has caused it to sink out of view. The play received such lukewarm responses that Stoppard completely revised it, removing or drastically condensing the lengthy scientific explanations in the dialogue for the 1994 run at Lincoln Center, where it was more successful.[103] There is now very little science left in the play. Yet, as Stoppard readily admits, "it's not the physics that's the problem . . . it's the story, the plot, the narrative, the mechanism, the twins, all that."[104]

There are many good analyses of the play, and there is no need to repeat them here. For our purposes, it is helpful to consider two key points about *Hapgood*. As John Fleming notes, "The structure of *Hapgood*, like that of other Stoppard works, is intimately related to its content."[105] The main motifs of the play, such as wave-particle duality, twins and doubles, the bridges of Königsberg, the double-slit experiment, and quantum theory as a metaphor for human nature, are all gathered under the overarching theme of Heisenberg's uncertainty principle. Even the play's structure enacts this idea, since "nearly every one of the twelve scenes has a double."[106] This structure of doubling "helps reinforce Stoppard's central thematic concern," which is to explore the duality of personalities by using the uncertainty principle as a metaphor.[107] The play also clearly reflects and muses on the methodology of science. *Hapgood*, indeed, "embodies the form of a scientific paper: 'Act I leads to a hypothesis; act 2 carries out the experiment. The denouement leaves to us the interpretation of the results.' "[108] The main finding or result—the answer—is given up front, at the start of the play, just as in a scientific paper.

Conclusion

Most of the plays discussed in this chapter are well known—*Arcadia*, *In the Matter of . . .* , and *The Physicists*, for example. Others, like $E = mc^2$, are barely known today. That they have been so long neglected or obscure is perhaps due to their topicality, but it may also be due to the way in which they utilize the stage and the concept of theater that they embrace. Their emphasis on theatricality—particularly that which induces action as well as reflection—made them unfashionable in a New Critical paradigm, but all that has changed. As Natalie Crohn Schmitt writes, "The view that the interest of theater is not in the stories told—or at least not in them alone—but rather in the moments they facilitate has had a profound effect on theater history, allowing works not highly regarded as literature to be reclaimed for the (redefined) canon."[109] In addition, the chapter has focused in part on documentary science plays, which draw on actual historical figures and sources and lay claim to authenticity and verisimilitude through faithful and often verbatim use of this source material. Yet we have seen how such claims are questionable; subtle shifts of emphasis occur in the manipulation of the material. Are these documentary science plays more "real" and authentic than those like *Galileo* and *Copenhagen* that use historical scientists to fictional ends? In chapter 8, such issues will feature prominently as I discuss the implications of—and debates about—staging science and history.

Having established the ways in which physics was represented by and engaged in some earlier plays, we turn now to looking at one play about physics and physicists—and so much else—whose success on stage and in print merits a chapter of its own. Michael Frayn's *Copenhagen* is for many synonymous with science on stage, and it has shown itself to be a remarkably enduring piece of theater not just for its own qualities but for the controversies that have sprung up around it.

4 *Copenhagen* Interpretations: The Epistemology of Intention

••••••••••••••••••••••••••••••••••••••

In many respects, the play *Copenhagen* is the culmination of themes we have encountered in so many science plays discussed so far. Three historical figures meet in the afterlife to relive a moment that had a defining impact on them personally, as well as on the course of science and, arguably, of history. Those figures are the Danish physicist Niels Bohr, his wife, Margrethe, and the German physicist Werner Heisenberg, and the moment they are trying to live again is Heisenberg's visit to the Bohrs in occupied Denmark in the autumn of 1941. Bohr and Margrethe exchanged pleasantries with Heisenberg for a few minutes inside their bugged home; then the two men went for a walk in order to speak more freely to one another. They returned only a few minutes later, Bohr extremely upset, and Heisenberg made a hasty exit. From then on their friendship cooled, yet neither man ever revealed what exactly had been said during that brief walk. Frayn revisits this decisive moment in history and in science, positing three "drafts," as the characters call them, each with different outcomes, and the audience essentially has to choose which draft it prefers, since no concrete answers are explicitly given in the text, and since the characters' own memories of the events consistently fail them or show themselves to be flawed or subconsciously edited and revised.

Frayn was amazed by the public enthusiasm for the play. As he commented, "When I wrote it, I didn't expect anyone would perform it, let alone come and see it."[1] He was reluctantly prepared to offer it as a radio play if no one would stage it, simply to get it performed at all.[2]

The play is remarkable for the degree to which it engages "real" science. The characters repeatedly utilize "thought experiments" to test their hypotheses and explain their scientific ideas. Even the structure of the play, divided into three "drafts," mimics a scientific paper: "So, Heisenberg, why did you come? . . . Tell us once again. Another draft of the paper. And this time we shall get it right. This time we shall understand," says Bohr toward the end of act 1.[3]

However, these efforts are foiled by the inadequacies of human memory. The audience watches the characters in *Copenhagen* enact a process of conscious, effortful recall of a transforming moment. This moment has a certain resonance with readers of Proust, since it is much like the episode in *Remembrance of Things Past* in which Marcel's tasting and smelling of the madeleine dipped in tea transports him back to his childhood in Combray.[4] In *Copenhagen*, however, the central moment is experienced not by one but by three characters, which complicates things immensely. We quickly see that each remembers the meeting differently, down to the seemingly concrete facts such as time and place:

> MARGRETHE: You couldn't even agree where you'd walked that night.
> HEISENBERG: Where we walked? Fælled Park, of course. Where we went so often in the old days.
> MARGRETHE: But Fælled Park is behind the Institute, four kilometres away from where we live!
> HEISENBERG: I can see the drift of autumn leaves under the street-lamps next to the bandstand.
> BOHR: Yes, because you remember it as October!
> MARGRETHE: And it was September.
> BOHR: No fallen leaves!
> MARGRETHE: And it was 1941. No street-lamps![5]

The characters cannot agree on when or where the meeting occurred, let alone what words were exchanged. In conspicuously postmodern mode, the play calls into question the reliability of memory and the notion of any absolute truth, suggesting that our memories are governed and shaped by an unconscious process of editing and revision. The main device by which Frayn conveys this idea is the genre of theater itself: the staging of the play reinforces the idea of the elusiveness of facts.

Observing the Observed

As a script for performance, the text of *Copenhagen* offers no hints as to staging because there are no extradialogic stage directions. This makes reading the play difficult but is liberating for a director. In the Broadway production, there were just three chairs on stage and no other scenery or props; the stage itself was round and atomlike, and the characters orbited within it as they paced through the three drafts. Director Michael Blakemore made the stage into yet another metaphor, as some of the audience

Production shot from *Copenhagen*. Photo by Conrad Blakemore.

Production shot from *Copenhagen*. Photo by Conrad Blakemore.

sat in a tribunal at the back of the stage, watching and "judging" the action in stark marble stalls. They were in turn watched by the rest of the audience. Clearly, Frayn does not let the audience forget the implications of this mysterious event for both history and science; science itself, as much as the individual characters, is on trial before us. Yet he pointedly withholds a definitive "draft" that would solve the mystery. There is no comforting finality, only more troubling questions.

The Danish director Peter Langdal and his actors in Copenhagen, Denmark, also discovered how important the bare stage is to the play, and how that element is in fact its essence. Frightened at first of so many words, so much difficult physics, they began rehearsals by building "a copy of Niels Bohr's home," complete with a dining room, a piano, "that whole scenery filled up." Then something extraordinary happened: "For, I think, the first two weeks we walked around in this very normal house, and then after a while I took away all of the furniture, and it ended up like it is now. With three chairs. It's so simple. And I thought if we had continued one more month we wouldn't have had any chairs at all! I don't know how long a time we should have been working to have no actors!"[6]

The minimalism Langdal arrived at is essential to the play, but there is a misconception about it that needs correcting. The fact that the play is so visually spare and so textually dense does not mean that it makes for poor theater or that it is unsuited to the stage. Contrary to John Lahr's assertion in his review of the play in the *New Yorker, Copenhagen* did not start out as a radio play. Lahr claims that "in Frayn's hands, the Broadway-show shop is turned into a talking shop, and more's the pity. He had originally imagined 'Copenhagen' as a radio play, and if you shut your eyes, all of the play's meaning is clear."[7] This is exactly the idea expressed by physicist Jeremy Bernstein in his description of a staged reading of the play in Cambridge, England as part of the Jesus College colloquium on *Copenhagen*: "This format of just reading the play is I think very good. One can concentrate on what it says without being distracted by theatrical business."[8] Many readers of the play adhere to this view, seeing the text as sufficient in itself.

Such remarks indicate that there is a need for a better understanding of the connection between form and content that informs the whole structure of the play and invests it with much of its meaning. Far from being merely "distracting," the "theatrical business" of the play is central to its thematic purposes. And far from being better suited to radio than to the stage, the play demonstrates a "thoroughgoing connection between form and content."[9] Perhaps we need a better understanding of theater on the part of scientists; the bridge between the two cultures cannot be one-way.

The event that Bohr, Margrethe, and Heisenberg are trying to recall was far from ordinary; its historical significance was of tremendous proportions, affecting the development and use of atomic weapons and the political map of Europe, not to mention the history of science itself. And one would think that its being a shared memory, a communal experience (much like theater itself), would increase its accessibility as well as its reliability. When you know you are experiencing a supremely important moment, you are surely going to remember it accurately. Yet, as Frayn's play shows, this seems to make no difference; it is equally difficult to retrieve, no matter how hard the characters concentrate on establishing the truth of what happened. In *Copenhagen* you have interminable arguing over what happened, despite (or because of) three witnesses to the event.

This is entirely different from the kind of internal, private memory that we each experience individually and that is captured in the famous madeleine episode during the narrated recollections in Proust. The observer is also the participant. Can he both observe and experience simultaneously? According to Proust, yes. But Frayn employs Heisenberg's uncertainty principle to show us why this cannot be. "You can never know everything about the whereabouts of a particle, or anything else," Heisenberg says in *Copenhagen*, "because we can't observe it without introducing some new element into the situation, a molecule of water vapour for it to hit, or a piece of light—things which have an energy of their own, and which therefore have an effect on what they hit."[10] Further on Bohr explains:

> [Einstein] shows that measurement—measurement, on which the whole possibility of science depends—measurement is not an impersonal event that occurs with impartial universality. It's a human act, carried out from a specific point of view in time and space, from the one particular viewpoint of a possible observer. Then, here in Copenhagen in those three years in the mid-twenties we discover that there is no precisely determinable objective universe. That the universe exists only as a series of approximations. Only within the limits determined by our relationship with it. Only through the understanding lodged inside the human head.[11]

Similar explanations of this idea are sprinkled throughout the dialogue, yet Frayn avoids oversimplifying it through easy metaphor. He reminds us in his substantial postscript to the play that the uncertainty principle "as introduced by Heisenberg into quantum mechanics was precise and

technical. It didn't suggest that everything about the behaviour of particles was unknowable, or hazy. What it limited was the simultaneous measurement of 'canonically conjugate variables,' such as position and momentum, or energy and time. The more precisely you measure one
variable, it said, the less precise your measurement of the related variable
can be; and this ratio, the uncertainty relationship, is itself precisely for-
mulable."[12] Frayn warns us that "the concept of uncertainty is one of those
scientific notions that has become common coinage, and generalised to
the point of losing much of its original meaning."[13] But Frayn makes his
thematic connection absolutely clear: "What the uncertainty of thoughts
does have in common with the uncertainty of particles is that the difficulty is not just a practical one, but a systematic limitation which cannot
even in theory be circumvented," namely, that "thoughts and intentions,
even one's own—perhaps one's own most of all—remain shifting and
elusive. *There is not one single thought or intention of any sort that can ever
be precisely established.*"[14]

Demonstrating the Science Anecdotally

As the three characters work through the possible scenarios of the meeting, they literally enact this idea through Frayn's brilliant merging of
theme with form. The dialogue brims with vivid demonstrations of the
applicability of the uncertainty principle to the workings of memory, and
as they talk the actors orbit the stage like the electrons, neutrons, and
protons they signify. As soon as we become certain about one of them,
we are made to doubt another, and so on. Again and again the uncertainty
principle is both explained anecdotally by the characters and simultaneously enacted by their movements. The first example of this comes in act
1 when the two men recall how differently they skied, Heisenberg shoot-
ing down the hill and Bohr taking his time. Frayn turns this into a demonstration of uncertainty:

> HEISENBERG: Your ski-ing was like your science. What were you waiting
>   for? Me and Weizäcker to come back and suggest some slight change of
>   emphasis?
> BOHR: Probably.
> HEISENBERG: You were doing seventeen drafts of each slalom?
> MARGRETHE: And without me there to type them out.
> BOHR: At least I knew where I was. At the speed you were going you were
>   up against the uncertainty relationship. If you knew where you were

when you were down you didn't know how fast you'd got there. If you knew how fast you'd been going you didn't know you were down.

HEISENBERG: I certainly didn't stop to think about it.[15]

This skiing analogy not only conveys the scientific idea but also gives a sense of the characters of the two men. It is a deft and charming way of enlightening us without lecturing. A similar demonstration of a basic scientific idea comes in the beginning of act 2, as the characters recall the scientist Kramers, one of Bohr's "boy wonders" at the Institute when Heisenberg first arrived there. In discussing him they explain the structure of an atom to the audience:

HEISENBERG: All the rest of us had to work in the general study hall. Kramers had the private office next to yours, like the electron on the inmost orbit around the nucleus. And he didn't think much of my physics. He insisted you could explain everything about the atom by classical mechanics.

BOHR: Well, he was wrong.

MARGRETHE: And very soon the private office was vacant.

BOHR: And there was another electron on the inmost orbit.

HEISENBERG: Yes, and for three years we lived inside the atom.

BOHR: With other electrons on the outer orbits around us all over Europe.[16]

One of the play's most memorable anecdotes immediately follows this lesson in atomic structure, and it too demonstrates the uncertainty principle. It is the "papal progress" tale, referring to the earlier designation of Bohr as the pope of physics, and it is here quoted in full:

BOHR: You remember when Goudsmit and Uhlenbeck did spin?

HEISENBERG: There's this one last variable in the quantum state of the atom that no one can make sense of. The last hurdle . . .

BOHR: And these two crazy Dutchmen go back to a ridiculous idea that electrons can spin in different ways.

HEISENBERG: And of course the first thing that everyone wants to know is, What line is Copenhagen going to take?

BOHR: I'm on my way to Leiden, as it happens.

HEISENBERG: And it turns into a papal progress! The train stops on the way at Hamburg . . .

BOHR: Pauli and Stern are waiting on the platform to ask me what I think about spin.

HEISENBERG: You tell them it's wrong.

BOHR: No, I tell them it's very . . .

HEISENBERG: Interesting.

BOHR: I think that is precisely the word I choose.

HEISENBERG: Then the train pulls into Leiden.

BOHR: And I'm met at the barrier by Einstein and Ehrenfest. And I change my mind because Einstein—Einstein, you see?—I'm the Pope—he's God—because Einstein has made a relativistic analysis, and it resolves all my doubts.

HEISENBERG: Meanwhile I'm standing in for Max Born at Göttingen, so you make a detour there on your way home.

BOHR: And you and Jordan meet me at the station.

HEISENBERG: Same question: what do you think of spin?

BOHR: And when the train stops at Berlin there's Pauli on the platform.

HEISENBERG: Wolfgang Pauli, who never gets out of bed if he can possibly avoid it . . .

BOHR: And who's already met me once at Hamburg on the journey out . . .

HEISENBERG: He's travelled all the way from Hamburg to Berlin purely in order to see you for the second time round . . .

BOHR: And find out how my ideas on spin have developed en route.

HEISENBERG: Oh, those years! Those amazing years! Those three short years![17]

This train journey has several functions. First, it conveys the excitement of science and the passion it sparks for the scientists, as was briefly discussed earlier (see chapter 2). Second, Frayn manages to drop the names of an astounding number of important physicists, establishing his credentials in writing about this subject, and also giving a sense that even though the main focus is on these two physicists, they were always part of—and responsible to—a larger community, a common "culture" of scientists. Finally, the description of Bohr's movement via train and the individual scientists attempting to pin him down as he makes his journey sounds a lot like the attempt of the scientist to measure the exact position of a particle in its trajectory—with equally dubious results, as we know from the uncertainty principle. One can never precisely know what Bohr was thinking, or where the atom is, and like the atom his "position" changes each time it is measured.

Perhaps the best explanation of the uncertainty principle comes in Heisenberg's monologue in act 2 about the streetlights in the park:

And that's when I did uncertainty. Walking round Fælled Park on my own one horrible raw February night. It's very late, and as soon as I've turned off into the park I'm completely alone in the darkness. I start to think about what you'd see, if you could train a telescope on me from the mountains of

Norway. You'd see me by the streetlamps on the Blegdamsvej, then nothing as I vanished into the darkness, then another glimpse of me as I passed the lamp-post in front of the bandstand. And that's what we see in the cloud chamber. Not a continuous track but a series of glimpses—a series of collisions between the passing electron and various molecules of water vapour. . . . Or think of you, on your great papal progress to Leiden in 1925. What did Margrethe see of that, at home here in Copenhagen? A picture postcard from Hamburg, perhaps. Then one from Leiden. One from Göttingen. One from Berlin. Because what we see in the cloud chamber are not even the collisions themselves, but the water-droplets that condense around them, as big as cities around a traveller—no, vastly bigger still, relatively—complete countries—Germany . . . Holland . . . Germany again. There is no track, there are no precise addresses; only a vague list of countries visited. I don't know why we hadn't thought of it before, except that we were too busy arguing to think at all.[18]

Heisenberg's "Eureka!" moment is described in a way that brilliantly brings in allusions to previous anecdotes—the papal progress, the walks in the park—so that they cumulatively reinforce the science. Frayn shows the audience the connections and how they collectively demonstrate the uncertainty principle.

As these examples and many others in the text show, Frayn does not begin his play with a definition of the uncertainty principle but lets the audience come to it through a layering of variations on the theme, illustrations conveyed in the form of anecdotes that demonstrate the science using the characters as the subjects. Frayn lets the anecdotes define the ideas before giving them their scientific labels, which is a much less condescending method than lecturing the audience—and much more dramatically effective. Far from merely telling the audience what the science means, the dialogue performs it because the textual definitions are reinforced visually by the way the actors circle the stage and interact with each other. The culmination of this occurs toward the end of the play. "Copenhagen is an atom. Margrethe is its nucleus. About right, the scale? Ten thousand to one?" says Heisenberg. "Now, Bohr's an electron. He's wandering about the city somewhere in the darkness, no one knows where. He's here, he's there, he's everywhere and nowhere. Up in Fælled Park, down at Carlsberg. Passing City Hall, out by the harbour. I'm a photon. A quantum of light. I'm despatched into the darkness to find Bohr. And I succeed, because I manage to collide with him. . . . But what's happened? Look—he's been slowed down, he's been deflected! He's no longer doing exactly what he was so maddeningly doing when I walked into him!"[19] Bohr protests: "But, Heisenberg, Heisenberg! You also have been deflected! If people can see what's happened to you, to their piece of light, then they can work out what must have happened to me! The trouble is knowing what's happened to you! Because to understand how people see you we have to treat you not just as a particle, but as a wave. I have to use not only your particle mechanics, I have to use the Schröd-inger wave function."[20]

Thus, before we know it, Frayn has introduced not only uncertainty but complementarity: "Particles are things, complete in themselves. Waves are disturbances in something else. . . . They're either one thing or the other. They can't be both. We have to choose one way of seeing them or the other. But as soon as we do we can't know everything about them."[21] Heisenberg further anthropomorphizes particles:

> And off he goes into orbit again. Incidentally exemplifying yet another application of complementarity. Exactly where you go as you ramble around is of course completely determined by your genes and the various physical forces acting on you. But it's also completely determined by your own en-tirely inscrutable whims from one moment to the next. So we can't com-pletely understand your behaviour without seeing it both ways at once, and that's impossible. Which means that your extraordinary peregrinations are not fully objective aspects of the universe. They exist only partially, through the efforts of me or Margrethe, as our minds shift endlessly back and forth between the two approaches.[22]

Explanations of other scientific ideas abound as well, not just of the uncer-tainty principle. Frayn covers fission, Schrödinger's wave function, ura-nium 235, the calculations for critical mass; he even deftly explains what cadmium is, by analogy with Bohr's fatherly and restraining hand on Heisenberg ("You always needed me there to slow you down a little. Your own walking lump of cadmium").[23] Toward the end of the play he even manages to provide a brief history of science itself in one breathtaking page of text, contextualizing the achievements of Bohr and Heisenberg within the framework of scientific progress since Protagoras:

> Not to exaggerate, but we turned the world inside out! Yes, listen, now it comes, now it comes. . . . We put man back at the centre of the universe. Throughout history we keep finding ourselves displaced. We keep exiling ourselves to the periphery of things. First we turn ourselves into a mere adjunct of God's unknowable purposes, tiny figures kneeling in the great cathedral of creation. And no sooner have we recovered ourselves in the Renaissance, no sooner has man become, as Protagoras proclaimed him,

the measure of all things, than we're pushed aside again by the products of our own reasoning! We're dwarfed again as physicists build the great new cathedrals for us to wonder at—the laws of classical mechanics that predate us from the beginning of eternity, that will survive us to eternity's end, that exist whether we exist or not. Until we come to the beginning of the twentieth century, and we're suddenly forced to rise from our knees again . . . with Einstein.[24]

As usual, Margrethe brings the dialogue back to earth and to specifics:

MARGRETHE: So this man you've put at the centre of the universe—is it you, or is it Heisenberg?

BOHR: Now, now, my love.

MARGRETHE: Yes, but it makes a difference.

BOHR: Either of us. Both of us. Yourself. All of us.

MARGRETHE: If it's Heisenberg at the centre of the universe, then the one bit of the universe that he can't see is Heisenberg.

HEISENBERG: So . . .

MARGRETHE: So it's no good asking him why he came to Copenhagen in 1941. He doesn't know! . . . I've typed it out often enough. If you're doing something you have to concentrate on you can't also be thinking about doing it, and if you're thinking about doing it then you can't actually be doing it. Yes?[25]

This in turn brings them back to the analogy of skiing: "Swerve left, swerve right, or think about it and die."[26] All this is carefully laid out to show that no matter how hard they may try, these characters will never ascertain what actually happened during that visit. They can only "look back and make a guess, just like the rest of us. Only a worse guess, because you didn't see yourself doing it, and we did."[27]

Margrethe gives an uncharacteristically long speech about the "personal" nature of all this science: "Because everything *is* personal! . . . I'm sorry, but you [Bohr] want to make everything seem heroically abstract and logical. . . . [But] it's confusion and rage and jealousy and tears and no one knowing what things mean or which way they're going to go."[28] Science is done by real people, experiencing the same things as the rest of us, including a process of mistakes, blindness, hesitation, and disagreement. And all this is good and natural, to be not hidden but accepted. Perhaps this is what Margrethe means when, a few lines later, she refers to the Copenhagen interpretation as an attempt to "re-establish humanism"—putting the individual back at the center of the universe means dealing with the emotions that both distinguish us from the animals but also make us so "messy," so immeasurable and unpredictable.

Having established the connection between particles and humans—having used one to demonstrate how the other works—Frayn goes on to reinforce this analogy for the rest of the play. "Complementarity, once again," says Heisenberg. "I'm your enemy; I'm also your friend. I'm a danger to mankind; I'm also your guest. I'm a particle; I'm also a wave. We have one set of obligations to the world in general, and we have other sets, never to be reconciled, to our fellow-countrymen, to our neighbours, to our friends, to our family, to our children. We have to go through not two slits at the same time but twenty-two."[29] This dilemma, seemingly impossible, is exemplified further in the deeply emotional monologue Heisenberg delivers toward the end of the play, which begins with his comment about "a strange new quantum ethics."[30] This last phrase has been understood by some critics as a call for a new morality, but Frayn emphatically dismisses such an interpretation.[31] The speech hints instead at Frayn's real preoccupation as philosopher. It is a paradox that the demonized Heisenberg, as Bohr says, "never managed to contribute to the death of one single solitary person in all your life," whereas the beloved Bohr by his own admission played a "small but helpful part in the deaths of a hundred thousand people."[32]

By the end of the play our own certainties have shifted; we have moved from an initial sympathy with Bohr to an ambiguity about him and his motives and a burgeoning empathy with Heisenberg in his morally complex, difficult situation. This aspect of the play has sparked an ongoing debate. Frayn has succeeded in questioning the received opinion about Heisenberg, which is one of condemnation; as fellow physicist Lise Meitner wrote, "Heisenberg and many millions with him should be forced to see these camps [concentration camps] and the martyred people. His appearance in Denmark in 1941 is unforgivable."[33] *Copenhagen*'s Heisenberg shifts the spotlight away from the ethics of his visit to its motives, from external judgment to inner intentions.

Startlingly, Frayn finally gives us a moment that seems to defy the uncertainty principle, when the three characters perform their third and final "draft" in the imaginary living room of the Bohrs:

> HEISENBERG: . . . Here I am at the centre of the universe, and yet all I can see are two smiles that don't belong to me. . . . I can feel a third smile in the room, very close to me. Could it be the one I suddenly see for a moment in the mirror there? And is the awkward stranger wearing it in any way connected with this presence that I can feel in the room? This all-enveloping, unobserved presence?

MARGRETHE: I watch the two smiles in the room, one awkward and ingrati-
ating, the other rapidly fading from incautious warmth to bare polite-
ness. There's also a third smile in the room, I know, unchangingly cour-
teous, I hope, and unchangingly guarded.

BOHR: . . . I glance at Margrethe, and for a moment I see what she can
see and I can't—myself, and the smile vanishing from my face as poor
Heisenberg blunders on.

HEISENBERG: I look at the two of them looking at me, and for a moment I
see the third person in the room as clearly as I see them. Their importu-
nate guest, stumbling from one crass and unwelcome thoughtfulness to
the next.

BOHR: I look at him looking at me, anxiously, pleadingly, urging me back
to the old days, and I see what he sees. And yes—now it comes, now it
comes—there's someone missing from the room. He sees me. He sees
Margrethe. He doesn't see himself.

HEISENBERG: Two thousand million people in the world, and the one who
has to decide their fate is the only one who's always hidden from me.[34]

The disembodied smiles captured in the mirror seem to allow the characters a distance they had never had before, and thus a fleeting insight as
they can both see themselves and be seen, be in motion and yet measur-
able, do and think at the same time. But it is only fleeting. Frayn's reiterated point is that if self-knowledge is flawed, how much more limited
is our access to other people's thoughts and motives. In the Proustian
(modernist) view, the memory of what happened during the 1941 meeting
is like those smiles in the mirror; it can be retrieved only through the
unexpected tapping of deeply sealed recesses of our unconscious, through
the mechanisms of involuntary memory. The great difference in Frayn's
plural, postmodern conceptualization of memory, of course, is that there
are not really any sealed vessels anymore—there is ultimately no one
"true" memory, only multiple "drafts."[35] I will explore this idea later in
the discussion of Theatre de Complicite's *Mnemonic*.

The Role of Performance

It might seem paradoxical to claim that Frayn's play depends on performance for the illustration of its ideas when the foregoing discussion of
the play's anecdotal demonstrations of the uncertainty principle and complementarity has relied so heavily on textual analysis. But that is precisely

the point: to show the interdependence of form and content, the performativity inherent in the speech acts of the dialogue, which makes the characters actually do what they are talking about as they move about the stage. Scientists who have discussed the play have focused on its text; few have cited the importance of the collaboration between Frayn and the director Michael Blakemore

Blakemore's own invaluable explanation of how the play works in performance appeared in the *New York Times* in April 2000—a transcript of a talk he had given at a symposium on the play at the City University of New York, attended by (among others) hundreds of scientists: "There are a number of walks that the characters take in the play. Of course, there is only a certain distance you can travel on a stage unless the motion is circular. But if it *is* a circle, you can go on a walk forever. I felt that if we had the actors moving rather like particles within an atom, there would be times when this could be instructive and other times when as a metaphor it might be quite interesting."[36] This is the clearest statement of the play's performativity, and it shows how thoroughly organic the form is to the substance of the play, not merely an arbitrary choice of genre. But what is most interesting is how Blakemore extends these ideas to a metacritical look at what the play has to say about theater itself:

> One of the reasons it [the play] works with an audience is that the actual act of going to the theater and seeing a play supports a lot of the propositions in *Copenhagen*. Putting on a play itself is a sort of scientific experiment. . . . This is not a naturalistic play. We're not trying to pretend that what we're seeing is *real*. The audience must listen to the arguments, empathize with the characters' emotions, and create the reality for themselves. One of the lovely things about the play is that it shows us what the theater can do that none of the other media can. The play could not exist in its present from as a film. It couldn't be television. It is entirely a play. And it's very encouraging in that sense because it suggests a direction in which the theater might go. . . . It is a pleasure to see how an audience responds, for instance, to the notion of complementarity . . . [and] how the principle of complementarity exists in our everyday lives. . . . This is something that *Copenhagen*, in fact, deals with. . . . Throughout *Copenhagen*, it was extraordinary the way the act of theatergoing supports the various concepts in the play.[37]

Blakemore here defines precisely what has distinguished plays like *Copenhagen* and *Arcadia* from more routine works that employ science. They manage to use the theater on at least two levels: as a place for engaging science for the exploration of philosophical ideas, and as a way of investigating the nature of performance, the "act of theatergoing" itself.

The Bohr Letters: Life Imitates Art

Copenhagen continues to stir debate and controversy, as well as some sensational revelations. In February 2002, due to the attention the play has brought to the relationship between Bohr and Heisenberg and particularly the play's sympathetic, almost heroic depiction of Heisenberg, the Bohr family released letters it had previously had sealed in the Bohr archives. This epistolary cache (accessible on the Web site of the Niels Bohr Institute: www.nbi.ku.dk) contains drafts of letters Bohr wrote to Heisenberg but never sent alluding to their meeting in 1941, as well as letters from Heisenberg to Bohr. Then, in November 2002, at a colloquium on *Copenhagen* at Jesus College, Cambridge, Frayn revealed the existence of a letter written by Heisenberg to his wife, Elisabeth, during his weeklong visit to Copenhagen in 1941. The full text of this letter was published on the Internet and in the *New York Review of Books* in August 2003.

Collectively, these newly revealed letters shed fascinating light on the event at the center of the play. Heisenberg's letter, written on Tuesday, Thursday, and Saturday evenings, reveals that the meeting actually took place over several days and at various venues, including Bohr's house and the Institute. This gives a very different picture of the two men; rather than meeting briefly, having a devastating blowup, and never speaking again of their clash, they did in fact reconcile their differences to a certain extent and continue to meet both throughout that weeklong visit and afterward. As Thomas Powers puts it: "But when did Bohr grow angry, and just how angry did he get? Was the breach immediate and deep? The new letter suggests not."[38] Indeed, on the last of the three evenings they were together during Heisenberg's week in Copenhagen, an "especially nice" time was had, with Bohr "reading aloud" and Heisenberg playing Mozart's A-major sonata.[39]

What impact does this new letter then have on the play? The discovery that the meeting actually took place over several days rather than on a single evening seems trivial in itself, as all it throws into question is the unity of time and place that Frayn employs; it hardly alters the main thematic points of the play. In fact, Heisenberg's letter only confirms how aptly Frayn has captured the concerns of the two men and turned them into fascinating epistemological questions. Heisenberg tells his wife, "The conversation quickly turned to the human concerns and unhappy events of these times; about the human affairs the consensus is given; in questions of politics I find it difficult that even a great man like Bohr can not separate out thinking, feeling, and hating entirely." Given Heisenberg's dangerous position and the fact that the letter would almost certainly be censored before it reached his wife, this language is naturally ambiguous. But less ambiguously he mentions further on "the unavoidable political conversations, where it naturally and automatically became my assigned part to defend our system."[40] This admission of acting, of playing a role, of supporting the Nazis only in an official capacity, is surprisingly bold. It is also completely consistent with the depiction of Heisenberg in Frayn's play.

The Bohr Institute's letters show Bohr speculating much as the play does about what Heisenberg *thought* that Bohr was thinking and vice versa. "I got a completely different impression of the visit than the one you have described," writes Bohr to Heisenberg in one draft of a letter never sent. "I am greatly amazed to see how much your memory has deceived you," he states in another. Altogether there are eleven documents, and except for one written by Heisenberg to Bohr, all are unfinished drafts written by Bohr in the late 1950s and early 1960s, addressed to Heisenberg but never sent. True to Bohr's obsessive draft-writing habits they are all drafts of essentially the same letter, with slight (but significant) changes of wording here and there.

The letters show how divergent the two men's recollections of the events were, and Bohr lays claim to the stronger, more accurate memory recall. In draft after draft, Bohr continues his effort to revisit the scene and substance of the famous meeting. It should be noted that the tone of these letters is generally cordial and warm, as well as diplomatic. But one letter (Document 5a) contains a decidedly different tone: Bohr seems less certain of his recollection, admitting the difficulty of forming "accurate impressions" of events that had so many participants. As if to bear this out, the date he recollects for the meeting is inaccurate, as he himself seems aware: "1942(?)."

Bohr dictated some of his memories of wartime conversations with German physicists to Margrethe, including his meeting with Heisenberg. Document 6 is in Margrethe Bohr's handwriting, and in it Bohr describes how he kept silent after Heisenberg stated his conviction that the war, if it did not end with a German victory, would be decided by means of atomic weapons. The crux of their misunderstanding lies in their interpretations of Bohr's silence. Heisenberg mistook the silence as shock that atomic bombs could possibly be made so soon. But Bohr says in these letters that his silence was due to his awareness that "a great matter for mankind was at issue in which, despite our personal friendship, we had to be regarded as representatives of two sides engaged in mortal combat." In Document 7, also in Margrethe's hand, Bohr reiterates what he had

said in the previous draft about the two men having such different interpretations of the same event. He mentions his "cautious position" and
maintains that his memory is "quite definite" about what was said.

Document 10 is one of the most important new letters. Here, Bohr
takes exception to Heisenberg's claim to have come to Copenhagen to
tell Bohr that the German physicists would try to prevent the use of
atomic science for weapons of mass destruction. This is directly relevant
to Frayn's play, in which Heisenberg expresses moral qualms about developing atomic weapons and suggests that he deliberately impeded the German war project by failing to do the necessary calculations.

As these brief examples show, the new documents are interesting for
several reasons. First, in terms of Bohr's own memory, the letters show
what Bohr recalled many years after the meeting. At times he is very
certain of his memory, but at others he doubts himself; overall, what he
claims to have recalled provides remarkable confirmation of what modern
neuroscience has discovered about the unreliability of memory. We cannot rely on the "real" Bohr now any more than we can the Bohr in the
play. Also, the letters show disagreement over the basics, such as where
the meeting took place. This is consistent with Frayn's play, and the question is still not resolved about whether the meeting was in Bohr's house
or in his office or on a walk. Finally, the letters reveal an intense preoccupation with getting the precise wording Bohr wanted, stemming from his
deep awareness of this event as a public moment of historical significance.
The letters eerily corroborate Frayn's whole notion of the three dead figures revisiting this incident, rather than letting spirits rest—Bohr keeps
reviving this moment, trying to pinpoint what exactly happened. In this
respect the new letters support Frayn's contention that we can never
know the thoughts and intentions of others, or even ourselves; as he has
recently and emphatically stated, "The epistemology of intention is what
the play is about!"[41]

Much attention has been given to what the new letters reveal about
substantive issues, such as politics, science, and history. But for all the
debate, there are no real revelations, and nothing that radically challenges
Frayn's depictions of the two characters. Heisenberg's letter, conscious as
it is of the Nazis reading over his shoulder, alludes in the vaguest terms
to "purely human concerns" and "political reasons." It does not resolve
the mystery of why Heisenberg came to Copenhagen. Bohr assumed that
Heisenberg had been authorized by the German government to discuss
military secrets, but his letters reveal the new information that Heisenberg came to Bohr independently, acting on his own initiative, and risking

his neck to do so. If anything this seems consistent with Frayn's nuanced depiction of Heisenberg.

How have people with firsthand knowledge of the events reacted to these new letters? Physicists Hans Bethe and Klaus Gottstein agree that the letters do not clarify anything about the visit. Summing up the reaction of the majority of those with firsthand experience of the people and events involved who have read the newly released documents, Gottstein writes that the letters "do not, in an unambiguous way, solve the enigma of what happened during the critical brief discussion between Bohr and Heisenberg in 1941." Gottstein surmises, as have many others, that Heisenberg simply came to visit Bohr with the naive assumption that "Bohr would be ready, as he always had been in earlier times, to discuss with him possible solutions for complicated problems."[42] This is akin to what Peter Langdal proposes as Heisenberg's motive. The Danish director recalls how the actor playing Bohr arrived at a very simple and moving explanation for why Heisenberg came to Copenhagen: "Henning said, 'I think Heisenberg came to Copenhagen because he opened up a door to something he was scared about, and then he needed to come back to his father figure, who also opened up the same door, so they could hold hands.' Yes, hold hands. And that's what theater is all about, hold hands; trying to know yourself by knowing each other."[43]

In addition to suggesting a simple human motive for Heisenberg's visit, Langdal speaks powerfully here about the necessity of theater, not just its entertainment value; he reinforces Frayn's point, quoted earlier (and expressed at this same seminar), that theater performs a valuable function in society by engaging two groups of people—the actors and the audience—in a confrontation, a silent dialogue. This seems to be utterly consistent with the "third culture" dialogue envisioned by C. P. Snow, as quoted in chapter 2. Langdal also says that "actually we built our whole production on that one sentence," which he notes is "one of the best explanations of the question I've got actually, and it came not from a physicist but from an actor."[44] The hypothesis that Heisenberg just wanted to come to Denmark to "hold Bohr's hand" in this difficult time is both deeply moving and wonderfully simple. It also echoes the play itself, for in several places the phrase is used: in act 1, for example, Heisenberg says three times to Bohr, "you held out your hand to us"; "you held out your hand to us"; "you held out your hand to us then, and we took it."[45] For all the complexity of the scientific ideas and the seemingly unsolvable mystery at the heart of the play, *Copenhagen* is like all great dramas in this respect: the motives of the characters can be described in such straightforward and universal terms, in one brief sentence.

These Bohr letters prompted Frayn to write the additional "post-postscript" to the play in which he calls the release of materials "the most surprising result of the debate set off by the production of the play"—an unprecedented instance of life imitating art, one might say. But as Frayn himself points out, these letters by no means resolve the question of what really happened in that meeting in 1941. Their only real effect on the play is to call into question Frayn's sympathetic rendering of Heisenberg as having moral qualms about developing an atomic bomb. Frayn acknowledged the impact of these revelations on his work, but only in terms of the postscript, not the play itself: "When and if we do another edition of the play, I think I will certainly have to record [in the postscript] that these letters have been published, and we now have Bohr's direct testimony as to what his feelings were."[46] It is interesting that the Bohr family thought that by releasing these documents they would lay to rest the questions surrounding the 1941 meeting. If anything, the letters—and the newly unearthed Heisenberg letter—only keep fueling the debate. Most recently, the publication in Germany in 1995 of Rainer Karlsch's book *Hitlers Bombe* has added to the controversy by its sensational claim that the German nuclear weapons program was far more extensive and advanced than previously thought and that a group led by Kurt Diebner carried out an experiment with a nuclear device in the spring of 1945. Whether or not this claim can be sufficiently substantiated, Karlsch's book has already had a significant impact; it expands the focus on Heisenberg as the sole leader of the German effort to develop nuclear weapons outward to other German groups, and suggests that—despite some of their later claims to the contrary—German scientists were indeed working to create a bomb for Hitler. It will be interesting to see how this new development affects the continuing debate about the precise nature of Heisenberg's role in the German nuclear weapons program, and how that in turn colors the reception of *Copenhagen*.[47]

One might argue that Frayn's play, with its theme that we can never be certain of either our own or other people's thoughts and motives, will always be protected no matter how many documents may emerge attesting to "the truth" of the event, since its central concern is not finally historical but philosophical and epistemological in nature. Even if new revelations continue to bring us closer to the truth of what really happened in Copenhagen in 1941, the play's themes—and its harnessing of the science to convey them—will remain relevant. "Whatever was said at the meeting, and whatever Heisenberg's intentions were, there is something profoundly characteristic of the difficulties in human relationships, and profoundly painful, in that picture of the two ageing men . . . puzzling for all those long years over the few brief moments that had clouded if not ended their friendship. It's what their shades do in my play, of course. At least in the play they get together to work it out."[48]

5 Evolution in Performance: The Natural Sciences on Stage
• • • • • • • • • • • • • • • • • • •

> Inventing new forms that assert themselves an instant and then
> disappear or transform themselves is the work of evolution, and
> also that of a certain kind of theater. Darwin: link between
> biologist and director. Pay homage to the development of living
> forms and detect in the work of the scientist—Darwin—the
> exploits of a man, the work of his imagination.[1]

As we have seen, in the long tradition of science on stage, the field of
physics has dominated the genre. The natural sciences made spo-
radic appearances in works such as Shaw's *The Doctor's Dilemma* (biochemistry) and Ibsen's *An Enemy of the People* (epidemiology and public
health), then later *Inherit the Wind* (the Scopes trial pitting evolution
against creationist theory). But recently, biology, anatomy, and other biological sciences have begun to appear more frequently in plays. This chapter will look at plays that engage evolutionary theory, particularly Timberlake Wertenbaker's *After Darwin*, and also at a spate of plays addressing
problems and issues raised by research in genetics and cloning, such as
Caryl Churchill's *A Number* and Shelagh Stephenson's *An Experiment
with an Air-Pump*. The collaborative theatrical work of Jean-François
Peyret and Alain Prochiantz in their Darwin-inspired trilogy will also be
mentioned but is given full treatment in the final chapter in the context
of new "alternative" science plays.

One may account for the surge in new plays about the natural sciences
by the increasingly vexed and complex ethical issues that they address—
a reflection of our collective societal concern with where science is leading us as we become capable of manipulating our bodies and our surroundings in new ways. But beyond issues there are also personalities. Every
field needs its representatives; just as Galileo and Einstein form the human
face of so many physics plays, Darwin proves to be something of a poster

boy for plays about the natural sciences, although as Peyret points out this is mainly in the English-speaking world. Darwin's legacy continues to furnish dramatic interest and cultural debate. At a time when the Grand Canyon is being festooned with biblical plaques claiming that it is a mere two thousand years old and was created by God, it is not surprising that plays about Darwin and evolution are proliferating and proving to be of relevance, interest, and perhaps even unexpected urgency.

Evolutionary Theory at Play

Inherit the Wind (1955) was quite popular on stage, in print, and on film. In striking fashion the play borrows from Ibsen's *An Enemy of the People*, also enjoying renewed popularity at the same time in Arthur Miller's 1957 version, which overtly connected the play with McCarthyism. Both plays center on a perceived public enemy (Cates and Stockmann, respectively) who is put on trial. Cates's crime is teaching Darwin's theory of evolution in the public school system. He endures a formal trial with jury, judge, and witnesses; Stockmann faces a no less daunting tribunal of the enraged townspeople who gather to hear him in the fourth act of *Enemy*. The public enemy is better educated and informed than the common people, and they castigate him for it. They also use the device of forbidding the discussion of the science itself: in *Inherit the Wind*, Cates barely speaks, and the lawyer Drummond is forbidden by the judge to call his fifteen scientific witnesses to explain evolutionary theory, while in *Enemy of the People* Stockmann is allowed to speak publicly only if he does not mention his medical/scientific findings. Yet each one finds a way around this attempted silencing through skillful public oratory (Stockmann directly, Cates through his eloquent lawyer), driven by a conviction that he is right and in the moral minority.

Both Cates and Stockmann are officially defeated, confirming the latter's idea that new concepts take a generation to become accepted. Each overcomes this defeat by finding strength in personal relationships—Stockmann with his family, and Cates with Rachel—and in the formulation of idealistic new projects designed to improve humanity's lot, with the women as loyal assistants to help carry them out. Thus in each play popular ignorance officially wins the day, yet private integrity and the aura of these men as underdogs and pioneers clearly makes the audience side with them as the real winners. We know who is right, both in

the fictional world of Ibsen and in the "real" historical world of Lee and
Lawrence.

Ultimately both plays show that the real enemy of the people is
ignorance, and as the character of the thirteen-year-old boy Howard in
Inherit the Wind shows—a character with several counterparts in *Enemy
of the People*—the only hope is in education and in the next, more enlight-
ened, generation. This is made especially clear by having Howard be the
first character we see on stage, as he scares a young girl with his talk of
evolution. Yet he is reassuringly "normal," with his Norman Rockwell
overalls and fishing rod, his boyish utterances and his love of worms, and
his submission to having his hair matted down with spit to tame the
inevitable cowlicks—shades of Thornton Wilder's *Our Town*, and it is no
coincidence that Lawrence and Lee utilize such mainstream American
stereotypes for the same critical purposes.[2] Hallie Flanagan Davis also uses
this wholesome, gee-whizzing American boy figure ("Henry") at the center of her play $E = mc^2$. Henry is an ordinary, intelligent boy and, like
Howard, represents a kind of hope for the future—but only, the playwrights make clear, if the education of these and all other children is
allowed to continue.

In addition to the Ibsenic influence, *Inherit the Wind* has a Faustian
theme, as Cates is vilified for his departure from the scripture and seen as
an overreacher. It also reminds one of *Galileo* because at the beginning of
the play Rachel comes to the jail to beg Cates to recant—to renounce
his teaching of evolution and admit that he is wrong. Like Galileo, Cates
is human and wavers; unlike Galileo, but only at Drummond's gentle
urging, he remains staunch in his beliefs in court. *Inherit the Wind* has
other interesting Brechtian overtones in its theatricality. A sense of alienation arises from the presence of a cynical reporter, the one character
who speaks in verse, which certainly for the reader is incongruous and
alienating. There are many episodic scenes, and not much happens that
could be construed as private and individual—even the love scenes between Rachel and Cates have another, more public, purpose. The crowd
and the nonrealistic staging devices, such as having the courtroom blend
into the town (no wall to the back of the room), are trenchant reminders
of the highly public nature of the issues in the drama and of the evolution
versus creation trial as a performance, a highly metatheatrical experience,
almost a play within a play. The playwrights also avoid making Cates
the focus, and thus skirt the usual categories of hero and villain, because
Cates barely speaks compared with the two lawyers; he is merely an indi-

vidual caught in the clash of dialectical historical forces, the casualty of a paradigm shift. The play succeeds in keeping in focus the scientific, cultural, social, and political issues at stake rather than the lives of individual characters.

The World after Darwin

Timberlake Wertenbaker's play *After Darwin* (1998) also deals with the theory of evolution. It addresses such aspects as natural selection, mutation, adaptation, survival of the fittest, and extinction, and it also brings in more contemporary issues such as the role of technology in our lives (through cell phones, e-mail, and electronic games like Tamagotchi toys). The play succeeds in fusing these quite disparate topics and concerns. Whereas other plays like *Inherit the Wind* have addressed the impact of evolution, and it is clear that Darwin's ideas were constantly engaged in popular entertainments of the nineteenth century,[3] *After Darwin* is the first play to enact the ideas under discussion through a thoroughgoing integration of form and content.

Like *Copenhagen*, *After Darwin* is a good example of how contemporary science plays try truly to grapple with, engage, and elucidate the scientific ideas that they employ by building them into the structure of the play and performing them. Although not a stage or print success like *Copenhagen*, the play has some strong thematic points and represents an important if overlooked contribution to the genre of science dramas. It employs a split-setting technique similar to Stoppard's *Arcadia*, with the same actors doubling characters from two different historical periods, in Wertenbaker's case the mid–nineteenth century and the present. This is done to enhance the thematic use of evolutionary theory, a subject dealt with previously on stage only in *Inherit the Wind*, in which the science is discussed but not literally enacted as Wertenbaker attempts in *After Darwin*.

The present-day characters are actors, Tom and Ian, putting on a play about Charles Darwin (Tom) and Robert FitzRoy (Ian), the captain of the *Beagle*, who hired Darwin to accompany him as a naturalist on his voyages to South America. The scenes alternate between this historical costume drama and the present, in which Tom and Ian talk with each other and with the Bulgarian director, Millie, and the African American playwright, Lawrence, about the play, their private lives, their past experiences, and their present aspirations. Millie is trying to gain English citizenship by proving that she has "something unique to contribute."[4] Lawrence is a black professor whose success is due to the sheer survival instincts of his mother, who raised him in a ghetto in Washington, D.C., to read "Shakespeare, Milton, *Moby Dick*, . . . [but] no black writers. No writing on slavery . . . [and] it worked, I guess."[5] As one reviewer notes, Lawrence "has evolved and adapted more than anybody else onstage," although the desperate Millie is "trying to camouflage her identity and join the English species."[6]

As the play progresses, the action in each time frame becomes increasingly dramatic. Darwin and FitzRoy become estranged as the very religious captain feels increasingly threatened by the implications of Darwin's scientific findings and ideas. Meanwhile, Tom confides in Ian that his dream of becoming a film actor is about to come true; he has been hired to appear in a movie but must quit the play to do so, which means the play will have to be canceled. To save the production, Ian betrays Tom and sabotages his career by secretly e-mailing the film director and telling him that the homosexual Tom is HIV-positive.

Reviewing the play for the *New York Times*, Benedict Nightingale acclaimed it as "even more absorbing" than Wertenbaker's highly successful breakthrough play, *Our Country's Good*, and noted the similarities between *After Darwin* and *Copenhagen* in their energetic use of science on stage. The play presents evolutionary theory through two prisms: the scientific and the social. Darwin and FitzRoy confront the scientific theory and its philosophical and religious implications, while the modern-day characters act out social Darwinism, which seems to be defined as people being incredibly selfish in order to survive and having no moral qualms about their deeds so long as they can justify their motives in terms of sheer survival. Tom defends his defection to the film project by citing adaptation and survival; Ian justifies his betrayal of Tom as survival not just of himself but of the others in the production and of the play itself: "I don't want another two years without work. I want to survive, I want Millie to survive, I want this to survive."[7] The doubling of roles also draws attention to Wertenbaker's thematic emphasis and the parallels she is drawing: Darwin is played by Tom, and FitzRoy by Ian. FitzRoy became so distraught at the implications of Darwin's findings that he confronted his former friend with a pistol, a scene that is included in the play and that parallels Ian's distress and desperation to salvage the play at all costs. Just as FitzRoy wants his faith to survive intact, Ian wants the play to go on, yet both know that Darwin/Tom's decisions are essentially unavoidable and necessary and must be accepted. They object to the way, in their eyes, Darwin and Tom "play God," yet they fail to see their own interventions in the same hubristic light.

A subplot relates directly to evolutionary theory while also deepening the character of FitzRoy. In the historical scenes, references are made to FitzRoy's well-known penchant for converting "natives" to English dress, manners, culture, and above all religion. The prime example of this is Jemmy Button, "civilized" by FitzRoy and then reintroduced to native culture with disastrous results.[8] Wertenbaker's strategy of having Lawrence tell Jemmy's story dramatically highlights the idea of successful integration through adaptation. "[Jemmy] had adopted Englishness with total enthusiasm, but had then readopted the customs of his tribe with equal commitment, thus becoming perhaps one of the first people to suffer the stresses of biculturalism, a condition which was to reach epidemic proportions in the late twentieth century. Jemmy Button's own tribe is now extinct."[9] Has Lawrence's success at adapting to the dominant culture come at the price of denying his own roots? Jemmy's tragic story seems to cast Lawrence's tale of survival in an ambiguous light.

In addition to the ways in which the modern characters enact and thus confirm the ideas of evolution to which FitzRoy is so resistant, there is also an interesting technological angle to the play. Ian "babysits" a Tamagotchi toy for his daughter, and this virtual pet places on the actor persistent demands for virtual sustenance and attention. He must attend to its beeping interruptions, and do so speedily lest the creature die and he thus traumatize his daughter whom he sees only "every other weekend" and who will "never trust [him] again" if he lets the toy creature die.[10] The device of the Tamagotchi pet thus reinforces the theme of inhumanity and emotional depravity inherent in Ian's betrayal of Tom, and suggests that contemporary life is characterized by a greater "reality" in virtual relationships than in flesh-and-blood ones. It is a bleak message about the implications of technology as a dehumanizing force, a message further conveyed through Ian's use of e-mail to destroy Tom's plans and hopes.

After Darwin was first produced at the Hampstead Theatre, London, in July 1998.[11] Although the play is visually spare, the settings and the sound combined to intriguing thematic effect. For example, the Bach keyboard music for the show, opening the production and punctuating the scenes, is highly suggestive; the complex, sophisticated piano music points to our highest state of evolution, but it also reminds us that this is what Jemmy Button was tutored to aspire to, with disastrous results. The music—pinnacle of Western civilization—underscores the uncomfortable theme of "civilization" of "savages" like Jemmy Button, whose story punctuates the play and endows it with a postcolonial theme. Pointedly, Jemmy is absent from the stage and therefore has neither a voice nor a physical presence. Instead, the African American Lawrence is his surrogate and speaks for him about the taming and civilizing of the noble savage and the difficulty of "biculturalism."

Biculturalism, the way Lawrence tells it, is a kind of acting. You can adopt and shed identities in order to survive. In its own subtle manifestation of "conscious theatricality," the play conveys the idea that the best demonstration of evolution, specifically the idea of adaptation, is acting. As the play progresses, masks are stripped away: Millie has been acting, as we discover that she was not a famous director in Bulgaria but a cleaner in a theater there; this is actually her first time directing a show. Tom is acting in two senses, both on stage and in his deception of his colleagues. Lawrence inhabits two very different cultural spheres through role-playing, just as Jemmy had done. Acculturation becomes synonymous with acting, and it is shown as the only way successfully to adapt to, and master, one's environment. Here again, as with plays from *Doctor Faustus* and *The Alchemist* to *Copenhagen* and *Arcadia*, we see a science play self-consciously examining theater as a thematic motif, as well as the chosen genre or packaging for the author's ideas.

Since each of the characters in turn does some kind of acting in order simply to survive, let alone get ahead, the theater becomes a metaphor for evolution, as well as the other way around. The theater demands adaptation, not just as you prepare each character, but during each performance and throughout the whole run. It is a process of continuous flux and change, whether on a tiny scale of adapations very few would ever notice or a radical "mutation" into some completely different version of the character because the actor is reacting to input from the outside world, such as critical reviews or the director's notes. In *After Darwin*, for example, Darwin is supposed to imitate the calls of birds and small mammals for FitzRoy in act 1, scene 8. The actor playing Darwin/Tom rendered the animal sounds beautifully; he did not deliver them dispassionately, but lingered over them. This has a complex effect. First, the audience is aware of a complete discrepancy between Tom and the character he is playing, because he convincingly portrays Darwin as caring deeply and passionately about such things as animal sounds, while Tom has clearly shown that he could not care less about Darwin or very much else, for that matter; he seems utterly superficial, yet here he adapts to the requirements of professional survival. In other words, here is an enactment of the idea of adaptation, conveyed through the metaphor of acting. Second, metatheatricality is heightened because the audience is aware that it is an actor (playing an actor named Tom who is in turn playing Darwin) who is making these expert sounds, and in admiring the actor's ability to do this, we are brought outside the world of the play and the play within it, an act of alienation, brief but important.

In a similar way, the play's title indicates two consciousnesses at work simultaneously. The title *After Darwin* acknowledges the paradigmatic shift caused by the theory of evolution, not just in the sciences but in the whole ethos of modern life: everything has changed after Darwin, and "the contemporary world is one in which the abbreviation A.D. might stand more appropriately for After Darwin."[12] But it also refers specifically to a line in the dialogue: "You cannot be tragic after Darwin. . . . A new species of modern sadness, perhaps"—but not tragic.[13] This idea underpins the entire play and is maintained by the author of the play-within-the-play, Lawrence, who has adapted and survived to become successful, but it is challenged in the end by the character who "is on the side of the losers . . . a good man who gets it wrong, that is tragic."[14] The contention that there is no tragedy after Darwin invokes the notion of paradigmatic shifts not just in science but in the history of ideas. We *think* differently after Darwin. Death, disappointment, grief, loss—all are simply side effects and symptoms of our constant evolution. Or are they? In an interesting twist, this idea is echoed in the play's final moments when this same character again says he is "the one who gets it wrong. Means well. Does ill. Always. Tragic."[15] Ultimately, tragedy *is* shown to be possible, directly contradicting the notion that all has fundamentally changed—and for the worse—after Darwin.

What exactly Wertenbaker means by this rather vague assertion about tragedy is not fully articulated by the play. The statement hangs in the air like an epigram, but one more glib than pithy. What is its context? And why "tragedy"? Peter Morton has studied the influence of Darwin on the literary imagination and summarizes the arguments about which of the three genres—comedy, farce, or tragedy—most closely resembles evolutionary theory. "Some have asserted that Darwin's self-recognised plight of being jammed between his personal liberal optimism and the ferocious logic of Darwinism generated in others a sort of rueful mirth. . . . Perhaps comedy is the literary form which most vividly reflects the reversals in evolutionary thought, since randomness is as vital a constituent of comedy as determinism is of tragedy."[16] Yet it is also possible that "the crazy disorder of pure farce" as represented by Lear, Carroll, and Wilde is the most "Darwinian literature" because farce "dissolves away nearly all formal artistic structure."[17] The third, more traditional view is that "the most relevant literary genre, the one most fitted in its conception of human destiny to transform into art the biologists' new image of man, is tragedy," in which "the malignancies of nature revealed in King Lear are just what might be expected of an organic universe."[18] This latter view has had powerful proponents over the intervening decades. Some even maintain that in addition to causing a "Darwinian blight" upon subsequent literature, On the Origin of Species itself is written like tragedy, culminating in a kind of patricide: the final ousting of the "Heavenly Father from the sphere of natural history."[19] This is not entirely accurate, however, since Darwin reinstated the term "Creator" in the concluding paragraph of the final edition of the book. Although his motives are ambiguous, this gesture at least acknowledges the possibility of God.

The phrase "after Darwin" thus seems to point to a particular stance on evolution, perhaps in relation to such categories. Wertenbaker seems skeptical about the legacy of Darwin's theory even while she is fascinated by its scientific possibilities. She seems to be asking, "What next?" Rather like the two documentary physics plays, Uranium 235 and $E = mc^2$, this play is very much addressed to the audience's conscience and reason, showing a problem but asking the audience to ponder the possible solutions and finally coming to understand that they have a huge responsibility in their hands: no less than the fate of the earth. Will we go down the good road or the bad? It is up to us, the play seems to suggest, and it does this especially well in its use of the "survival of the fittest" notion. While well suited to the natural world, evolutionary theory as applied to human behavior and motives (social Darwinism) becomes sinister and morally corrupt. Wertenbaker conveys this especially through Millie, a hybrid of concepts like survival and adaptation mitigated by the very human notions of tenderness and empathy—two concepts that seem to serve no evolutionary function, or to give humans an evolutionary advantage, yet clearly are vital to our moral survival. The effort to maintain a moral position in an immoral world (the one represented by Tom and Ian and the Tamagotchi) eventually breaks her; when she sees her production collapse, and with it her hopes of remaining in Britain, she curls up into a fetal position on the floor and prepares to "go extinct."[20] But she doesn't; true to the play's dictum, tragedy is averted, and a compromise of sorts ends the play. "The production ceases, but Millie and Ian find love, while Lawrence affirms that Darwinism's true legacy is 'empathy' (73) in that it connects mankind to all other creatures on earth."[21]

Millie reminds us that the role of women and issues of gender and culture are particularly interesting in relation to science and theater. Wertenbaker has talked about being "fascinated" by science and by women in science; she also mentions the "cultural revolution" science has brought about.[22] She has posited that women have an evolutionary tendency to create glass ceilings for themselves, and are unable to deal with hierarchies or to understand them at all. "Women have been (up to this point at least) historically ineffective."[23] The idea that theater is the perfect laboratory for exploring our relationship to history, as one experiences great consciousness in the theater, has permeated Wertenbaker's work, right through to her recent play *Galileo's Daughter* (2003). In an early interview she raised this issue of how women organize themselves nonhierarchically and how radically different it is from men's stratified social structures. Perhaps Millie in *After Darwin* (the play's only female character) is an example of this quality and its repercussions: she approaches her role as director in a nonhierarchical way that is radically different from the men's way of seeing authority, and they rebel against her. In the end she is overwhelmingly defeated and indeed subsumed by them. It is odd, then, that she finds love at the last minute—such a conventional, clichéd outcome. Millie becomes a microcosm of the historical ineffectiveness of women that Wertenbaker contends dogs our history, our science, and our theater.

The play's sometimes heavy-handed and contrived use of concepts like evolution, adaptation, survival, and extinction has led to the criticism that Wertenbaker "bangs away at her theme a bit relentlessly. With everything from Tamagotchi pets to Balkan history pressed into what becomes a running debate about Darwinism, you certainly get the feeling that a program-seller might at any moment leap onstage to announce her imminent sex change."[24] *After Darwin* may be less successful than *Copenhagen* in its performative demonstration of scientific ideas, and it does seem to commit Snow's sin of reductiveness. But "the dramatic brew is rich and mentally nourishing, embracing as it does questions of God and godlessness, determinism and free will, biology and ethics."[25] In addition, as Mark Berninger has noted, the play draws an explicit parallel between evolutionary theory and the possible extinction of the theater as an art form: "By using Darwin's theory as a metaphor, theater is shown as a threatened environment (threatened by financial problems and the power of the film industry). It is an all too small ecological niche for survival in a world dominated by cruel Darwinist and capitalist struggle."[26] Theater as Galápagos is perhaps not so risible an analogy.

It remains to be seen whether this play will enjoy eventual success through revisions and through its participation in the sustained development of science on stage, but to date it is the only play to integrate evolutionary theory both thematically and formally in a thoroughgoing and innovative way. In its literal enactment of evolutionary theory, *After Darwin* also manages to set us straight on the often misunderstood relationship between evolution and progress. *After Darwin* could be seen as an accurate representation of the theory of evolution in that the play does not confuse evolution with improvement—does not make it seem like the mere passage of time causes inevitable improvement in the human race, or that evolution is synonymous with things getting better and us getting stronger and smarter. According to this false and implicitly elitist (as well as anthropocentric) assumption, natural selection brings mankind ever forward in a linear, progressive fashion. The two time periods in the play accurately suggest that, far from getting better, we are simply adapting to changing environments. It is not a case of the old being surpassed by the new, the simple becoming ever more sophisticated. Tom and Ian are not "better" than their Victorian counterparts; in fact, Tom is shallow and superficial, ignorant of what Darwin wrote and represents, and while Ian earnestly wants to understand evolution, and chastises Tom for his lack of interest in the subject, he shows poor judgment in his underhanded and unscrupulous act against Tom. Yet Wertenbaker avoids simple nostalgia for a "better" time through the sobering story of Jemmy Button's enforced adaptation and FitzRoy's stringent and unbalanced ideas. This is why the play succeeds on so many levels.

Nick Ruddick criticizes Wertenbaker's moral line as being too conventional, as if the use of the two time periods is supposed to herald some new postrelativity ethics: "Though the play is an intriguing work, its resolution relies in the end too much on a traditional ethical absolutism to resolve satisfactorily the questions of uncertainty raised by the divided action."[27] However, it is a stretch to claim that *After Darwin* should be proposing some new kind of ethics simply because it uses the technique of alternating time periods, especially when that technique seems clearly to pertain to evolution, not quantum mechanics as in *Copenhagen*. The play manages to juggle its many ideas fairly well, on the one hand championing scientific progress and on the other warning of the "tragedy" it can lead to. As we will see, other recent biology-centered science plays also address this problem of whether there are ethical limits in the pursuit of knowledge, especially in relation to genetics.

Other new plays about Darwin and evolutionary theory have emerged; Snoo Wilson's *Darwin's Flood*, Jonathan Marc Sherman's *Evolution*, and Crispin Whittell's *Darwin in Malibu* are some of the more lighthearted examples. In addition, according to the journal *Interdisciplinary Science Reviews*, the late Stephen Jay Gould was planning to develop and perform a play about evolution: "[He] came up with a wonderful collage on evolution, from Darwin's books and letters. He intended to play Darwin himself and recite his monologue in Italian, as the Piccolo [theater] confirmed at the beginning of September 2001 when presenting its programme for the following year to the press. But by January, Gould was too ill to come to the rehearsals."[28] Darwin and evolutionary theory continue to provide rich biographical and metaphoric opportunities to playwrights. Most recently, the French neuroscientist Alain Prochiantz teamed up with the prominent theater director Jean-François Peyret to produce a play about Darwin entitled *Des Chimères en automne*, which premiered in Paris in late 2003. Another Darwin-based piece, *Les Variations Darwin*, was produced the following year. These will be considered in the conclusion to this book.

Genetics and the Ethics of Experimentation

A cluster of plays dealing with genetics has emerged, reflecting the general interest in—and prominence of—the new problems engendered by genetic research. Three of these new plays either were written entirely by scientists or had direct scientific collaboration, and are briefly mentioned here. First, there is Scottish playwright Tom McGrath's *Safe Delivery*, a play about gene therapy cowritten with his daughter, a geneticist, and sponsored by the Wellcome Trust. Next, scientist and painter Elizabeth Burns's *Autodestruct* is a play that links cloning with therapeutic strategies in cancer research as Burns dramatizes the repercussions and implications of a cancer patient cloning himself in order to find "the ultimate cure for cancer." Finally, the scientist Carl Djerassi's play *An Immaculate Misconception* deals with the intricacies of fertility in the modern age. All these plays are grounded in firsthand knowledge not just of the science they engage with but of the laboratory culture that produces the scientific results from which we all benefit.

The most theatrically interesting of the new group of genetics plays are Shelagh Stephenson's *An Experiment with an Air-Pump* and Caryl Churchill's one-act play *A Number*. Neither Stephenson nor Churchill is a scientist, but both are skilled playwrights. Together they explore new territory for the theater in their utilization of the scientific possibilities of genetics and cloning.

Like *After Darwin*, *Arcadia*, and *Oxygen*, *An Experiment with an Air-Pump* (1998, the same year as *Copenhagen* and *After Darwin*) alternates between two time periods. The scenes are set in the fin de siècle years of 1799 and 1999, showing scientists of very different times struggling with essentially the same moral dilemma about the uses and abuses of science. The playwright sets up these two periods as the bookends of modern science: 1799, the Enlightenment, as such questions were just arising, and 1999, where science is now, as the Human Genome Project is completed. It should launch a whole new age of progress and social improvement, but, as Stephenson implies, the mapping of the human genome gives rise to appalling dilemmas. The play seems to indicate that the great hope of science to help society, to be humane and progressive, to work toward social improvement, has become utterly corrupted by the enticements of industry, big business luring the best scientists away from pure, unattached research into nefarious exploitative purposes. Kipphardt and numerous other science playwrights present this theme as well.

This is a very dark play about science, the playwright having almost nothing good to say about it or its practitioners. Stephenson's scientists are incarnations of Faustus at his worst. In the 1799 plot, a totally corrupt and unscrupulous anatomist, Armstrong, tries to seduce a hunchbacked and vulnerable serving girl, Isobel, simply to have a closer look at her deformity. When she learns the truth and tries to hang herself in her despair, he finds her still breathing and finishes her off by smothering her—so eager is he to get his hands on her body for his research. He has earlier admitted to a relish that borders on sexual pleasure at dissecting deformed corpses, and confessed to waiting for disabled people to die so that he can examine their bodies: "We've got our eye on an undersized fellow, about three foot tall. He's not at all well. He'll not see out the winter." Armstrong's defense is simple: "Digging up corpses is necessary if we're to totter out of the Dark Ages."[29]

In the 1999 plot, we see the implications of such progress. The scientist is Ellen, who has been offered a lucrative position with a pharmaceutical company based on her groundbreaking, controversial work on genetics—fetal diagnostics, to be exact. She has some entertaining exchanges with the electrician, Phil, who is working on her house and likes to chat about the weird, "tricky stuff" of the paranormal, the pseudoscientific, *X-Files* stuff, the twentieth-century versions of necromancy and alchemy perhaps.[30] Yet for all his seemingly wacky ideas, Phil grasps the ethical dilemma of Ellen's work: "If they can map your genes before you're born, they'll soon be wanting a little plastic card with your DNA details on. And if it says anything dodgy, it'll be like you're credit blacked. And then imagine this, people'll say I can't have this kid because it'll never get a mortgage. I mean, that's bloody mad, that."[31] Stephenson seems to imply an inversion: hard science is actually much more dangerous than the UFOs and spontaneous combustion Phil talks about. Ellen's husband, Tom, an unemployed English lecturer who offers the humanist viewpoint, agrees: "In the hands of people who don't understand it properly, . . . it'll be something else to judge people by and discriminate with. What starts off as something with all the forces of good behind it, will be swallowed up by the market-place."[32] There is no question which of the two cultures has the moral high ground in this play. Tom may be washed up, redundant, but he is in the right, and his criticisms carry weight.

Two major issues in the play are the impossibility of reversing knowledge, which we have seen in so many science plays, and the unscrupulousness of scientists. There is a constant tension in the play between the idea of using science simply to understand the world and using it to change the world, with the latter bearing nefarious implications. The phrase "moral qualms" pops up frequently in both time periods. "I've never had a moral qualm in my life, and it would be death to science if I did," boasts Armstrong, adding that "discovery is neutral" and "ethics should be left to philosophers and priests."[33] While Tom criticizes scientists for doing all their experiments "in a vacuum,"[34] Ellen maintains that "you can't not pursue something. . . . Once you know something, you can't unknow it."[35] In the end, she takes the job, by implication probably ending her withering marriage (and thus her direct links with the moral qualms) and entering into her own Faustian bargain.

Stephenson's cast of characters includes Roget, he of thesaurus fame and here the industrious and likable assistant to the scientist Fenwick in the 1799 scenes. Roget busily categorizes words and their synonyms even as he fulfills his duties as secretary to a leading scientist. He is clearly applying scientific techniques to language. Along with Ellen's husband, Tom, Roget is the moral center of the play, confronting the ruthless Armstrong about his unscrupulous behavior toward Isobel and posing key questions to his employer about the pursuit of scientific inquiry. Through the synonym-spouting Roget, the redundant English don Tom, the poignantly hapless fledgling playwright (Fenwick's daughter), and the poetry-struck

Isobel, the play stacks the cards as heavily in favor of the humanist view-point as Shaw's *The Doctor's Dilemma* does in favor of Art.

Language takes on similarly heightened importance in Caryl Churchill's play *A Number*. This one-act piece, performed at the Royal Court Theatre in London in 2002, takes cloning as a means to explore the problem of identity. A father, Salter, has had his son Bernard cloned, not once but "a number" of times, so there are now "a number" of Bernards, mostly in their thirties. The play consists entirely of dialogue spoken by two actors, the father and the sons (all represented—with great versatility—by one actor). Bernard One (B1) clearly had a terrible upbringing, painfully neglected and maltreated by his father, and the clones are an attempt to assuage the guilt-ridden Salter. In the end we are told that one son kills another. It is a modern twist on the story of Cain and Abel, since in genetic terms they are the same person.

In any Churchill play we pay special attention to the use of language, particularly her unique overlapping dialogue technique, and this one is no different. Language in a sense supplants even staging in a play without scenery, with just a chair and two actors. Michael Billington points out that in this "engrossing spectacle," Churchill is not giving us "a debate on the ethics of cloning." Like Ibsen and Wertenbaker, like Marlowe and Brecht, the playwright poses numerous questions yet offers no easy solutions. It is "a challenging form of moral inquiry," writes Billington. The key question the play asks is "from what the essential core of self derives: from nature or nurture, genetic inheritance or environmental circumstance?"[36] *A Number* contains hardly any scientific detail about cloning, nor does it use an extended scientific metaphor. What is remarkable is the depth of engagement with the issues underpinning the play, and the fact that the play asks us to bring the science to it. Rather than provide lengthy explanations of cloning and genetics, *A Number* draws on our own understanding of these issues to fill in the blanks, as it were. There is an implicit assumption that we have at least some common base of knowledge about the science of cloning through news stories and general discourse. The success of the play confirms Churchill's assumption.

A Number in fact taps into a new vein in the nature-nurture discourse. Most audience members will approach the play with an awareness of this age-old question about which shapes our selves the most, our genetic makeup or our environment. The question still remains, but "the landscape and the climate have changed profoundly" because of the new scientific breakthroughs in understanding genetics as well as the social roots of human behavior.[37] At this point, "the mood in the nature camp is bullish and nowadays it is the one to claim the moral high ground."[38] However, the science is still in its infancy. "Much remains to be learned about how a cell containing a genetic recipe develops into an individual and there's something of a boom in this field, centred on the genes that control development. . . . As for the genes that underlie the things that make people interesting, like personality or intelligence, scientists have barely scratched the surface."[39] Finding the genetic basis for the "normal working of the mind" is the great challenge. Even if scientists do identify "a gene that affects mood here or intelligence there, each of these is only a cog in the most complicated machine we know. An inventory of cogs doesn't tell you how the machine works."[40] Churchill's play captures the strange mixture of being at once highly sophisticated yet completely un-knowledgeable: "It's an adventure isn't it and you're part of science," says B2, one of the clones.

> B1: Can we talk about what you did?
> SALTER: Yes of course. I'm not sure where what
> B1: about you sent me away and had this other one made from some bit of my body some
> SALTER: it didn't hurt you
> B1: what bit
> SALTER: I don't know what
> B1: not a limb, they clearly didn't take a limb like a starfish and grow
> SALTER: a speck
> B1: or half of me chopped through like a worm and grow the other
> SALTER: a scraping cells a speck a speck
> B1: a speck yet because we're talking that microscope world of giant blobs and globs
> SALTER: that's all
> B1: and they take this painless scrape this specky little cells of me and kept that and you threw the rest of me away
> SALTER: no
> B1: and had a new one made
> SALTER: no
> B1: yes
> SALTER: yes
> B1: yes[41]

Churchill's exploration of the nature versus nurture debate is totally current, as this is a highly visible debate right now, with new evidence on both sides. It is also germane to her feminist concerns as in earlier plays like *Top Girls*. *A Number* may lack female characters and overtly feminist themes, but in its conclusion that so much of our identity is not genetically predetermined but is a construct based on social norms and institutions, the play's message would seem to be very liberating for women.

It is interesting that all three recent plays analyzed here, involving the biological sciences in innovative and provocative ways on stage, are by women playwrights: Caryl Churchill, Shelagh Stephenson, and Timberlake Wertenbaker. In this regard, they inherit the mantle of Susan Glaspell, whose play *The Verge* already in the 1920s brought botany and genetics to the stage in the context of women's issues. Another notable new play in this regard is *Rosalind: A Question of Life*, by Deborah Gearing, which explores the often neglected contribution of Rosalind Franklin in the discovery of DNA. However, *After Darwin*, *An Experiment with an Air-Pump*, and *A Number* all go far beyond this area in their engagement of science. The plays integrate some of the key ideas in the modern biological sciences and dramatize the implications and issues they raise, issues the audience most likely confronts each time we open a newspaper or watch the news. The next chapter discusses plays that take us even closer to the everyday applications of science.

6

Mathematics and Thermodynamics in the Theater
●●●●●●●●●●●●●●●●●●●●●●●●●●●●●●●●

Mathematics is the sister, as well as the servant, of the arts and is touched by the same madness and genius.[1]

HAL: Some friends of mine are in this band. . . . They're all in the math department. They're really good. They have this great song—you'd like it—called "i"—lower-case I. They just stand there and don't play anything for three minutes.
CATHERINE: "Imaginary Number."
HAL: It's a math joke. You see why they're way down the bill.[2]

David Auburn's *Proof*, winner of the 2001 Pulitzer Prize, is generally known as the "math play." Yet this is quite a misnomer, since, as Karen Blansfield has noted, "*Proof* isn't about mathematics; it's about family connections, genius and insanity, the quest for love and the search for truth."[3] Strip away the few mathematical references in *Proof* (albeit some good math jokes) and you are left with a fairly conventional family melodrama that is mainly concerned with two traditional dramatic subjects: dysfunctional relationships (father-daughter, sisters, male-female lovers) and the (possibly hereditary) connection between genius and madness. Even with respect to the latter theme, the mathematics in the play has little to do with the issues under discussion; the mathematicians might just as well be painters or poets, since the central concern is not with mathematical genius specifically, but with any extraordinary creative or intellectual gift. The questions the play asks are likewise general ones: whether such genius leads inevitably to madness, and whether that madness is inevitably hereditary. Furthermore, the structure and staging of the play have nothing to do with its content; there is no formal innovation to go with the novelty of sprinkling some mathematical terms and some math jokes into the dialogue. In this respect, too, it falls outside the current study of science plays that are performative in nature.

But science plays generally are deeply indebted to *Proof*: its phenomenal success on stage and with the critics, and the fact that it has been made into a major film, may have done more than any other play apart from *Copenhagen* to bring about the current popularity of science plays. I want therefore to begin this chapter on mathematics in the theater with a brief look at *Proof*, why it is a great piece of theater if not a legitimate "science play," and why it is significant in this study.

The play takes place mainly on the porch outside the house of Robert, a brilliant but mentally disturbed mathematician, and his equally gifted daughter, Catherine. It opens with the two of them talking in the middle of the night of Catherine's twenty-fifth birthday, discussing among other things the development and nature of Robert's mental illness. We learn that Catherine is a loner with few friends who prefers mathematics to people, perhaps; has devoted herself to caring for her father, even sacrificing her education for him; and that they have a warmth and rapport through their shared love of numbers. Discussing how much time Catherine has lost because of her possible depression, they banter:

ROBERT: Call it thirty-three and a quarter days.

CATHERINE: Yes, all right.

ROBERT: You're kidding!

CATHERINE: No.

ROBERT: Amazing number!

CATHERINE: It's a depressing fucking number.

ROBERT: Catherine, if every day you say you've lost were a year, it would be a very interesting fucking number.

CATHERINE: Thirty-three and a quarter years is not interesting.

ROBERT: Stop it. You know exactly what I mean.

CATHERINE: (*Conceding*) 1729 weeks.

ROBERT: 1729. Great number. The smallest number expressible—

CATHERINE: —expressible as the sum of two cubes in two different ways.

ROBERT: 12 cubed plus 1 cubed equals 1729.

CATHERINE: And 10 cubed plus 9 cubed. Yes, we've got it, thank you.

ROBERT: You see? Even your depression is mathematical. Stop moping and get to work.[4]

This is a typical exchange between the father and daughter, establishing several key points: Catherine can hold her own in their mathematical discussions and is not awed by her father's brilliance because she matches it; the two know each other intimately and have a very strong bond; and, as in so many science plays, the passion for the ideas supersedes the personal problem or circumstance being shown or discussed.

But Auburn's great dramatic trick is to withhold from the audience a key piece of information: Robert is in fact dead, having passed away the week before, and this is the eve of his funeral. So his appearance in this opening scene is a hallucination, and we have been completely fooled. This fact is revealed to the audience only when one of Robert's former doctoral students, Hal, shows up and interrupts their "conversation." We find out that, with Catherine's permission, Hal has been going through the 103 notebooks that Robert left behind in the house, through of a sense of duty that "someone should read them."[5] This shock to the audience is a great dramatic device used extremely effectively here, and interestingly also used equally well in a contemporaneous story about a brilliant but disturbed mathematician: the film *A Beautiful Mind*, about John Nash. Nash suffered from hallucinations and heard voices constantly, just like Robert. In the film, two of his closest companions, his college roommate and a little girl he befriends in the park one day, are eventually revealed to be not flesh and blood at all but mere figments of his imagination. This is not revealed to us until well into the film, and it comes as a genuine shock. We have had time to get to know and like these characters and to regard them as real people in Nash's life, and to have to revise our conception of them so drastically requires a staggering mental leap.

Not only is this a great dramatic device, but it accomplishes the feat of so many of the science plays under consideration here in merging form and content—in this case, enacting or performing schizophrenia, *showing* the nature of bipolar personality disorder rather than merely telling about it. As we are getting over the shock in *Proof*, we learn through Hal how truly great a mathematician Robert was. Hal is no mathematical slouch himself, but he is realistic about the limitations of his ability. Catherine mistrusts him for this reason: she suspects his motives for going through her father's papers are not as altruistic as he claims and that he really wants to find something he can publish as his own. She tries to impress on him how crazy her father was: "He used to read all day. He kept demanding more and more books. I took them out of the library by the carload. We had hundreds upstairs. Then I realized he wasn't reading: he believed aliens were sending him messages through the Dewey decimal numbers on the library books. He was trying to work out the code."[6] Still Hal believes he will find some mathematical gem amid the deranged scribblings. Over the course of the play, he and Catherine fall in love, and she shows him a proof about prime numbers that has been locked up in one of the cabinets. It is written in the same kind of notebook as her father used, and in a very similar hand to his, but—and this is the second dramatic stunner of the play—Catherine reveals that it is in fact by her. "I didn't find it," says Catherine in the act 1 curtain speech. "I wrote it."[7] Hal and Catherine's sister Claire, who has arrived from New York for the funeral, immediately doubt this claim, and because we know about Catherine's own delicate mental state the audience too doubts it. We must figure out what to believe, and whose argument is most convincing. Could Catherine really be the author of a proof that reflects "some of the most important mathematics in the world," and that has eluded mathematicians "since there were mathematicians"?[8]

In the end, Hal overcomes his skepticism about Catherine's capabilities. He checks the proof with the help of his colleagues ("two different sets of guys, old geeks *and* young geeks") and concludes that it must be her work because "it uses a lot of newer mathematical techniques, things that were developed in the last decade. Elliptic curves. Modular forms. I think I learned more mathematics this week than I did in four years of grad school. . . . So the proof is very . . . hip."[9] The lovers are reconciled, and the play ends with Catherine staying on in Chicago (implicitly with Hal) rather than leaving with Claire, and with Hal and Catherine bent eagerly over the notebook absorbed not just in each other but in their shared passion for mathematics.

The play uses mathematical giftedness as a way of exploring the link between genius and insanity. Claire is in town for the funeral but also to sell the house and take Catherine back to New York to a mental health institution because she is convinced that Catherine has inherited their father's madness as well as his brilliance. The nature of the brilliance is not the point—Auburn could have used painting, writing, or any creative art as well as mathematics to illustrate this connection. To his credit, Auburn readily acknowledges the marginal role of the mathematics in his play, even while describing the pains he took to ensure that it was correct, such as seeking the advice of professional mathematicians and involving them in the rehearsal process. "I came to the mathematics pretty quickly, but it wasn't the first thing I started with. And I think the focus in the play remains on the kind of family drama, so that the math is more of a background. But it's a theme, you know, that runs through the play. . . . I figured out pretty quickly, first of all, that I could use math as a kind of metaphor for creativity in general, and also that the importance of beauty . . . the esthetics of mathematical work is a major component of it."[10] In the end, *Proof* succeeds in lifting the profile of mathematics with regard to the theatergoing public and even cleverly manages to challenge some of the very clichés it employs, such as the stereotypical math geek.

Given *Proof*'s success, one can only lament the lack of wider visibility of more deeply mathematical plays such as John Barrow's *Infinities*, staged in Italy and Spain to great acclaim but so far unperformed in English. There are in fact many recent "math plays" like *Partition* and *Incompleteness*, as well as *Possible Worlds*. Barrow's *Infinities*, Stoppard's *Arcadia*, and Theatre de Complicite's *Mnemonic* make up the trio of plays this chapter will explore because in very different ways, each work addresses the mathematical problem of time.

Our understanding of time and space has been influenced radically by modern physics, from Einstein's reconceptualization of the old Newtonian world to the more recent suggestions of multiple dimensions. Modern literature reflects this change, and one genre that has particularly seen a revolution in its handling of time and space has been drama. *Arcadia* (1993), *Infinities* (2002, 2003), and *Mnemonic* (1999) show especially why theater lends itself so well to playing with time. In particular they are concerned with the complex relationship between past and present, the concept of the future, and the theme of "the irreversibility of time."[11]

Arcadia contains an astonishing multiplicity of themes, ideas, and fields of knowledge, including physics, landscape architecture, and literary biography and criticism, all crammed into this one work and the "two hour traffic of the stage." Like *Copenhagen*, the play has generated a tremendous amount of interest from humanists and scientists over a broad range of disciplines, as well as general audiences. *Mnemonic* engages time in a different sense: memory and the latest findings in neuroscience that show us how shifting and unreliable memory is. This chapter will focus on the trio of plays *Arcadia*, *Mnemonic*, and *Infinities* with regard specifically to the use of mathematics, chaos theory, fractals, thermodynamics, and the physics of time and space. I have selected these three main plays as representative of how the theater engages such science-related topics as the neural processes involved in memory, the use of thought experiments in mathematics, and the deceptiveness of linearity and causality (and the Newtonian conception of time and space) in astonishingly original and theatrically provocative ways.

Much has already been written about *Arcadia*, especially in its handling of complex ideas, and my aim is not to repeat the excellent analyses that can be found elsewhere.[12] Rather, I will focus on the most overlooked aspect of *Arcadia*: the way in which the formal strategies of the play, specifically its "iterated plot structure,"[13] work in tandem with its thematic considerations, such as chaos theory and fractals. In performance the play becomes a working demonstration of the ideas at its core, and it demonstrates why the theatrical form is uniquely suited to the play's larger purpose; why it must be theater and no other genre. The emphasis here is on the necessity of performance, a feature of the play that—as in so many discussions of *Copenhagen*—often gets sidelined in the eagerness to dissect and explain the intellectual aspects of the work.

Temporal Tricks

Some special conditions apply to time in the theater. "Time works on the levels of real time, the time of the action, and theatrical time, which mediates between the other two in different ways according to the dramatic mode," writes theater scholar Marvin Carlson.[14] Bert O. States eloquently describes what makes theater special in its relation to time:

> The dramatist assigns his play to a scene, designated by language or by objects in space, without troubling to think how radically he has shifted the ground and conditions of our perception of the world. In a stroke he has altered our customary orientation to time and space. . . . The enacted play is the one art form, excepting the dance, that imitates human action in the medium, one might say, of human action. . . . It is precisely a purging: what is purged, at least on the level that concerns me here, is time—the menace of successiveness, of all life falling haphazardly through time into accident and repetition.[15]

Traditional plays have a specific use of structure, such as act and scene divisions and intermissions, which causes the audience to be "separated spatially and temporally from the performance" as well as from the actors. A play like Eugene O'Neill's *Long Day's Journey into Night* is in four acts, "with each opening and closing of the curtain between acts distinguishing audience-time from play-time. Act breaks used as intermissions call attention to this distinction even more strongly."[16]

By contrast, in some contemporary performance works, "no logical connections are provided; the work of interpretation and association is left to the audience, requiring more active participation than is called for traditionally and thus shifting the audience's attention away from anticipation to what John Cage calls the 'now-moment.'"[17] *Arcadia* is not so extreme, but it requires a great deal of mental input from the audience to make clear the implied connections. Like *Mnemonic*, it builds on what

Stoppard calls the "here and now" of theater in order to offer a systematic rethinking of how time works on stage.

When you are watching a film, you know that what you see has already taken place. Other pleasures may be had from film, but the presence of live actors and the kind of interchange that enables with the audience is not one of them. Nor does film allow for the unpredictability that comes with liveness; all sorts of events affect the performance and make it different each time, even if ever so slightly—an actor may forget a line, sneeze, miss an entrance, be feeling unwell that evening, or feel distracted, and he or she will likewise be affected by the presence of the audience members of whom all actors are of course aware.

This relates to a key idea behind both plays: chaos theory. The theatrical experience is like a chaotic system. Chaos, of course, is a misleading term, for as Valentine explains in *Arcadia*, it does not imply disorder at all, but an ordered, plottable system, albeit one whose course is affected by minute changes and disturbances, called "noise." The classic example of chaos theory in action is referred to by James Gleick in his popular book *Chaos*, in which he cites MIT meteorologist Edward Lorenz's 1972 talk entitled "Predictability: Does the Flap of a Butterfly's Wings in Brazil Set Off a Tornado in Texas?" The so-called butterfly effect has been used ever since to illustrate the central idea of chaos theory. The example of such a small system as the movement of a butterfly's wings being responsible for creating such a large and distant system as a tornado in Texas illustrates the impossibility of making predictions for complex systems; despite the fact that these are determined by underlying conditions, precisely what those conditions are can never be sufficiently articulated to allow long-range predictions.[18] Gleick's book was one of the main sources of inspiration for Stoppard's use of chaos theory in *Arcadia*. A theatrical production is just such a system. Although the word "production" sounds as if it is a material, permanent artifact like a painting or poem, since each performance of a given production of a play is different, there is no definitive artifact or product, no "final cut" as in film. The combination of live actors and live viewers affecting one another even while pretending not to interact (unless there is deliberate breaking of the fourth wall, and even then actors and onlookers also affect one another) renders the outcome of each given performance unpredictable. Small variations make the general outcome predictable (what the production will be like) but indeterminate in the exact permutations of each performance.

Stoppard links chaos theory to the essence of theater itself: its liveness, immediacy, and unpredictability. As his characters explain and enact chaos theory, and as the audience comes to understand and visualize it under their entertaining tutelage, the theatrical experience itself is providing a prime example of chaos theory. *Arcadia* shuttles between two different but interconnected time periods. To be exact, there are three: 1809 and 1812 and the present (1993), which complicates matters in interesting ways. The historical researchers in the modern period are trying to piece together what happened in the earlier period. They are often quite hilariously misled by clues that they misinterpret, and the audience meanwhile gets to see the actual events and the truth that is hidden from the modern characters. "By covering occurrences at three different moments in history on the stage . . . Stoppard offers the audience a scenario impossible outside the imaginary world: the exact description of events happening nearly 200 years apart. The interrelation of the past and the present together with the possibilities of interpreting or predicting either one thus form the central immediate concerns of the play."[19]

The Second Law of Thermodynamics

Stoppard reinforces Thomasina's observation that "you cannot stir things apart" several times throughout *Arcadia* through colorful analogies, much as Frayn repeatedly reinforces complementarity and uncertainty anecdotally in *Copenhagen*.[20] The audience must "get" the science without being lectured at. What we are learning in *Arcadia* is of course the second law of thermodynamics. According to the first law of thermodynamics, energy is conserved, which we all accept to be true. "The second law of thermodynamics takes on several forms. . . . [It] now reads 'the total entropy of any system plus that of its environment increases as a result of any natural process.' Entropy can be understood as the quantitative measure of disorder in a system."[21]

What does this actually imply? "The second law [of thermodynamics] gives us proof (if proof were needed) that we cannot in fact go back to or recapture the past: the pudding, having been stirred, cannot be unstirred. Chaos theory, the second of the play's scientific themes, goes still further in making the past inaccessible to us. Not only can we not step into the past, we cannot even discern its features through the noise of intervening subsequent events."[22] Even what looks like reliable, solid evidence from the past can be deceptive and open to misinterpretation. As we shall see, this occurs in *Mnemonic* as well.

Arcadia demonstrates chaos theory and the second law of thermodynamics in relation specifically to time in two main ways. First, Stoppard structures the alternating time periods in such a way as to confound the known laws of physics: the audience watches time reversed, the pudding unstirred, as the gradual revelations in each scene update and revise our understanding of both past and present. The very notion of a forward-moving, linear structure to life disintegrates; "time must needs run backward."[23] Second, Stoppard constantly poses the question of "what if" and shows that small choices lead to enormous changes. The most important example is in a later scene in which Thomasina, on the eve of her seventeenth birthday, asks Septimus to show her how to waltz—the latest dance craze, and the first dance to bring couples in an embrace. Clearly there is deep attraction between the two. But Septimus decides not to accompany Thomasina up to her room. The audience is inwardly screaming, "Go with her!" because Stoppard has *already revealed to us*—through Hannah and Bernard's dialogue—that Thomasina dies that very night in a fire caused by the candle in her room. Had Septimus been with her, that tragedy would probably have been averted, and this in turn, Stoppard implies, would have led to Thomasina going on to become a successful mathematician solving Fermat's last theorem and developing the revolutionary ideas that are mentioned in the play, thereby also freeing Septimus from devoting the rest of his life to shutting himself up as a hermit and trying in vain to solve Thomasina's unfinished theorem. This, in turn, would have had an impact on the other events in the play, for as Stoppard shows, everything is connected—all the disciplines and all the characters interrelate, no matter how remote they may seem to begin with.

Stoppard plays with time by having the audience witness the discrepancy between Bernard and Hannah's detective-like efforts at historical reconstruction and the actual events they are researching. Stoppard provides us with a whole new manifestation of dramatic irony, as we watch Bernard come to the wrong conclusions about what Lord Byron was doing at Sidley Park based on the evidence before him. Because we are actually watching the 1809 and 1812 scenes that Bernard is attempting to reconstruct, we know how to fill in the gaps, and we know he is drawing incorrect conclusions. This furnishes one of the play's most intellectually challenging aspects: we the audience have to juggle the tremendous amounts of information coming at us from both time periods and constantly add to and revise our understanding of the events being depicted and researched at the same time. Stoppard allows the audience to see both the future and the past that are being referred to by the characters, to watch them unfold and check them against what has already been learned or what will be learned. We are time travelers with privileged glimpses of what the characters themselves cannot know. And all the time, we know that this defies reality; it is simply not possible outside the theater.

Scenically these ideas are reinforced by the ingenious device of the "large table which occupies the centre of [the room]" and gradually accumulates props from both time periods that are not anachronistic but completely appropriate to both. Over the course of the play the table is home to an apple, a theodolite (a surveying instrument), a turtle, some books, large leather-bound hunting and gardening journals, "two geometrical solids," "a pot of dwarf dahlias," and many more items that work equally well in either the early nineteenth century or the present.[24] As Stoppard specifies, the table and indeed the room as a whole "should offend neither period."[25] This accumulation of objects allows two time periods, almost two hundred years apart, to coexist on the stage before the audience's eyes. It suggests a kind of time travel, as when Hannah drinks from Septimus's wine glass or Septimus peels and eats the apple offered to one of the characters in the modern period. It also underscores some of the ideas discussed in the play. For example, it reinforces Septimus's insistence on the importance of the "march" of progress, his rather conventional theory of cognitive development that hinges on the accumulation of knowledge: "We shed as we pick up, like travellers who must carry everything in their arms, and what we let fall will be picked up by those behind. The procession is very long and life is very short. We die on the march. But there is nothing outside the march so nothing can be lost to it. The missing plays of Sophocles will turn up piece by piece, or be written again in another language."[26] Chaos theory, of course, shows that this could not happen and that Septimus is wrong; the conditions that gave rise to Socrates and Euripides could never be exactly re-created, hence their works are forever lost.

The idea of the "march" of progress is borne out using the material resources of the theater in the simple device of the table accumulating the various items, each of which is invested with several semiotic meanings that can coexist. The turtle is both a pet and a paperweight. Its name is Lightning in one period, Plautus in the other. The apple is a symbol of the teacher in one period, an obvious reference to seduction in the other (and in both it represents knowledge and the loss of innocence). The theodolite is useful in one period, a quaint antique in the other (upstaged by Valentine's computer). The leather-bound books that pile up on the table are being written in by the earlier characters, and discussed and consulted as objects of study by the later ones.

What Stoppard has done is, in the eyes of some critics, to invent a whole new kind of theater, something akin to the modernist breakthrough in painting: "The fragmented story and plot lines in *Arcadia* rearrange sequence and causality just as a Cubist painting rearranges elements of reality into a new configuration in order to highlight underlying shapes often masked by habitual modes of perception. Stoppard here leaves behind the old 'Newtonian theater' of the well-made play."[27] He does so not simply by cleverly reshuffling scenes from both time periods but by intertwining the scientific and mathematical ideas of the play with its structure.

Fractals, for example, are not just discussed but are integral to the play's dramaturgy. "If you could stop every atom in its position and direction, and if your mind could comprehend all the actions thus suspended, then if you were really, really good at algebra you could write the formula for all the future; and although nobody can be so clever as to do it, the formula must exist just as if one could."[28] In saying this, Thomasina not only has foreseen fractals but has literally plotted the future in which Septimus will seclude himself in the single-minded attempt to find such a formula. The play's two time periods also beautifully convey its fractal ideas. Christopher Innes has shown that the dual setting of the play actually constitutes a fractal system, because although the play is divided fairly evenly between the two periods, the earlier scenes span a four-year period (1809– 13), and the present-day ones span just a few days.[29] This mimics fractal behavior. In *Arcadia* one group of characters wishes to know the future— to plot it mathematically—and another group seeks to know the past. Thomasina belongs to the former; Hannah and Bernard, the literary historians, belong to the latter. Their respective goals create a temporal tension in the play. It turns out that Thomasina was right about her speculations, but was too early for her time and died too young to pursue it (Valentine explains that she lacked the computer power to generate enough iterated algorithms to prove her idea)—it is the cruelest paradox that while she was ahead of her time, her own future was cut off. Meanwhile, we see that Bernard is wrong about what happened in the past, although he learns that much later (and to his great embarrassment).

In the last scene, after having the different time periods alternate throughout the play, Stoppard brings the two groups of characters together on stage. And there is another temporal trick: the modern charac-

Fractal. Copyright by Benoit Mandelbrot.

ters are dressed in Regency period costumes for a summer ball being held at Sidley Park, so the two periods are visually fused into one. Yet the characters are not aware of each other; they are in the same space, but not in the same time. The audience sees all the characters hovering around the same table that has been collecting the various artifacts from both periods; the actors seem to breathe the same air, almost to touch one another, yet they maintain their awareness only of the ones from the same time periods as themselves. This is quite ingenious, and invites delicious anachronisms such as Valentine looking like one of the 1809 characters yet tapping away at his computer. "Here, the two times have

The chaos game. Courtesy Michael Barnsley,
Australian National University, Canberra.

come so close together that their conversations and actions complement each other. . . . Especially at the end of the play, when in a powerful image two different couples of two different times waltz together through the same room, the audience is allowed to see" how brilliantly Stoppard collapses the distinctions between time past and time present.[30]

Stoppard has so effectively and subtly utilized staging to enact the scientific ideas and themes in the play that critics have often completely missed the connection. "I believe," writes one critic, "that the science in Stoppard's plays . . . [is] not intrinsic to the story. *Arcadia* could have been written without Fermat's last theorem, or even without chaos theory, as a literary lark around Lord Byron."[31] However, analysis of the way Stoppard integrates science into the structure of the play shows that scientific ideas are indeed intrinsic to the story. Speculations about what might happen if one tinkered with a work of literature in various ways, removing this or that character or event, or changing the outcome of the plot, violate the basic principles of literary criticism and provide further evidence of the need for analysis of science plays from a literary and dramaturgical perspective.

The impact of *Arcadia* on subsequent drama and theater has been extensive. The play showed how successfully one could incorporate sophisticated scientific ideas into the theater, and it spawned a rash of look-alikes in its wake—plays that use the same juxtaposition of different time periods (usually using the same actors doubling), with varying degrees of success. Timberlake Wertenbaker places the action in *After Darwin* alternately in the mid-1800s and the 1990s. As discussed in chapter 5, hers is among the most effective of the plays that use *Arcadia*'s dual-setting template because like Stoppard she merges this form with the content of the play. Shelagh Stephenson makes no attempt to do this in *An Experiment with an Air-Pump* with the result that her play seems derivative, her use of the dual setting merely incidental to the ideas in the play. Carl Djerassi and Roald Hoffman's *Oxygen* mimics *Arcadia* in its use of the dual time periods, and it engages a topic that is also featured in *An Experiment with an Air-Pump*—Lavoisier's discovery of oxygen or, more precisely, his discovery "that combustion is a process in which oxygen combines with another substance. Up till then they thought the combustible ingredient was something called phlogiston."[32] *Oxygen* dramatizes this finding in historical scenes that alternate with modern-day ones, but the dramaturgical strategy is not used to enact the science being described.

By contrast, as Lucy Melbourne notes, "Stoppard's emphatically nonlinear play requires its audience to constitute this aesthetic object both in prospect *and* in retrospect *at the same time*: the repetition and variation of plot components, when fed back in this way, gradually reveal the outline of *Arcadia's* underlying thematic structure." For example, the revelation of Thomasina's death by fire is strategically placed "between the rediscovery of her iterated algorithm and an explanation of her diagram of entropy, the idea that the world will eventually end through loss of heat." [33]

The title *Arcadia* signals another meaning to the notion of "heat death," although it may not be obvious at first and requires the kind of intellectual leap audiences have come to expect of Stoppard. The word has many associations, among them literary (Sidney's *Arcadia*, for example) and artistic (Poussin's painting *Berges d'Arcadie*, in the Louvre). "Arcadia was the Virgilian vision of a bucolic ideal, an ahistorical idealization born from the excesses of Republican Rome. To later generations it has become a place of contradictions, where life meets death and failure success," writes Stephen Abell. The painting

> shows three shepherds and a shepherdess standing before a tomb marked with the inscription "Et in Arcadia ego": "I too lived in Arcadia." It forms an understated revelation of existence tinged with extinction, of happiness circumscribed by sorrow, of paradise "qualified." The picture's strength lies in the quality of its enigma: the tomb has no name, the inscription no obvious speaker. Indeed, Poussin reworked the picture, removing a skull from its foreground. It now serves as an illustration of the precarious and ambiguous message that life can be both strengthened and threatened by the approach of destruction.[34]

The idea of Arcadia—particularly this interpretation of it, and not the Edenic one often ascribed to the term—thus relates to the second law of thermodynamics, one of the play's prominent motifs, and its implication that the universe is doomed. "So the Improved Newtonian Universe must cease and grow cold. Dear me," say Septimus at the end of the play, echoing (paradoxically in nonlinear fashion) what the modern-day character Valentine has explained earlier: "Heat goes to cold. It's a one-way street. Your tea will end up at room temperature. What's happening to your tea is happening to everything everywhere. The sun and the stars. It'll take a while but we're all going to end up at room temperature."[35]

Absorbing this unsettling information, Hannah responds by quoting Byron: "I had a dream which was not all a dream. / The bright sun was extinguished, and the stars / Did wander darkling in the eternal space."[36] These lines are from the opening of the poem "Darkness," Byron's "gothic extravaganza" in which he imagines "the effect on earth of the sun's being extinguished."[37] The poem's later lines are just as interesting and relevant, and worth quoting in their entirety to show how well they apply to the play's main themes:

> The world was void,
> The populous and the powerful—was a lump,
> Seasonless, herbless, treeless, manless, lifeless—
> A lump of death—a chaos of hard clay.
> The rivers, lakes, and ocean all stood still,
> And nothing stirred within their silent depths;
> Ships sailorless lay rotting on the sea,
> And their masts fell down piecemeal; as they dropp'd
> They slept on the abyss without a surge—
> The waves were dead; the tides were in their grave,
> The moon their mistress had expired before;
> The winds were withered in the stagnant air,
> And the clouds perish'd; Darkness had no need
> Of aid from them—She was the universe.[38]

Well before the second law of thermodynamics has been articulated, Byron lyrically captures the notion of the extinct universe, gone dark, cold, and lifeless after the death of the sun. This foreshadowing, implied by Stoppard's partial quotation of the poem, complements Thomasina's prescient use of mathematics. Stoppard brings these instances of foresight to life through the medium of theater, by juggling the two time periods and showing us that earlier ideas are confirmed by later discoveries.

Stoppard is known for "bringing alive conflicts between intellectual, artistic and moral propositions and the kinds of people who put them forward." In this he continues in the vein established by Shaw's brand of theater of ideas. According to Jeremy Treglown, "It's one of the paradoxes of Stoppard's work that while it often satirizes academics and biographers, no British dramatist since Shaw has been so concerned to teach, or to use theater as a medium for biography—or rather biographies, since Stoppard is above all a cultural historian."[39] Treglown calls Stoppard "the one-man Adult Education Centre."[40] He notes that Stoppard's choice of genre is quite deliberate, citing the playwright's dislike of narrative biography, which is a form Stoppard has "taken such trouble to attack that you wonder quite what he's fighting against."[41] There is a deliberate and direct relationship between choice of genre and choice of subject matter in the case of *Arcadia*, as in all of Stoppard's theater.

How to Do Things with Memories

Theatre de Complicite's *Mnemonic* juggles scenes from two stories and two interrelated time periods, but this is not another *Arcadia* look-alike. The two stories are linked through the central character, a young man named Virgil who is fascinated by the concepts of time, memory, and the past. Through him we learn about the Iceman discovered by mountaineers in 1991 in a glacier on the border between Austria and Italy, and the actor playing Virgil also plays the Iceman, or more accurately, his corpse as it is presented by the scientists intent on analyzing it. The other story revolves around Virgil's girlfriend, Alice, whose mother's death prompts her to embark on a sudden quest to find her father, a man she thought was dead. This journey takes her across Europe to Poland and Lithuania and on to Russia, and finally back to Poland again, into many people's lives as she tries—much like Bernard and Hannah in *Arcadia*—to track down this missing link from the past. Her story is told through phone calls with Virgil.

The word "mnemonic" means "of or designed to aid the memory." We talk about a mnemonic device, such as a rhyme or acronym that helps us remember something that is otherwise hard to do by rote. For example, a simple sentence we memorize in ninth-grade biology class, "King Peter came over from Germany Saturday," helps us recall the sequence "Kingdom Phylum Class Order Family Genus Species." As you would expect, *Mnemonic* is about memory, explored through the company's trademark approach: "communal storytelling that relied on the expressive powers of the body and the transforming capacity of inanimate objects."[42] The title points to a theatrical experience that is concerned with large-scale mnemonics—how things remind us of other things across cultures, epochs, geographies, and generations.

This play at first glance appears to bear all the hallmarks of science plays like *Copenhagen* and *After Darwin*, and to be more like these plays than the alternative science plays I will discuss later. True, there is a narrative that can be easily identified, and this narrative revolves around two characters who play out the familiar stuff of romantic drama, a young and impoverished couple trying to figure out their relationship. Contrapuntal to this story is the other narrative about the Iceman, which unfolds in alternating scenes to the main story and complements it, rather like *Arcadia* and *After Darwin*. Part of the play's brilliance lies in its ability to show how such seemingly disparate narratives relate to one another. But the play is also remarkable for the way in which it frames these two stories within a larger narrative: that of our own recollections, both private and cultural. The play makes us reflect on the making of memories, as well as on how the personal impinges on the public, the borders between them constantly shifting just the way memories themselves constantly shift and subtly revise themselves.

Neuroscience has relatively recently revealed this important insight into memory, using the latest techniques and experiments, and the play literally takes this as its starting point by opening with a lecture on the biochemistry of memory. This lecture is addressed directly to the audience, and Simon (Theatre de Complicite's cofounder, Simon McBurney) softens its didactic content by joking, swearing, using a broken chair in ingeniously suggestive ways to illustrate his points, and—most important—getting the audience involved by asking us to sit in the darkness with shades on our eyes and think back to our own childhoods.

The scientific information comes immediately, up front, while the audience's attention is at its best. "One of the last great mysteries is the one we carry inside our heads," Simon says, "how we remember, why we remember, what we remember."[43] He tells us that "human memory starts to degenerate when you are only twenty-eight years old," joking that "as I am now over forty the matter is becoming a little pressing."[44] Simon then launches into a substantial explanation of how memory works, according to the latest scientific findings. He explains that scientists used to believe memory retrieval worked like a computer: that "individual memories were stored and carried in individual brain cells," and to retrieve one your brain simply had to identify "the relevant brain cell, get into it and . . . wham . . . there was the memory . . . exactly the same each time. No variation."[45] But now we know that is not true, and that in fact, "when we remember it comes out slightly . . . different each time."[46]

Simon explains the idea of fragmentation, which is at the core of modern theories about memory:

> Different elements are, apparently, stored in different areas of the brain. And it is not so much the cells that are important in the act of memory, but the connections between the cells, the synapses, the synaptic connections. And these connections are being made and remade. Constantly. Even as I am talking to you part of your brain is changing. You are literally developing new connections between the neurons. They are being fabricated even as I speak. It's a process called sprouting. Think of that; you are all madly sprouting as I stand here, the biochemical ferment going on between the cells is unimaginable. And with the thousands of these connections being sprouted as I speak, we can think of memory as a pattern, a map.[47]

Having seized on this handy metaphor of the map of memory, Simon explains that it is not a neat, stable, "ordnance-survey" map but one that is "constantly changing and developing" so that "each time we read the map, thousands of roads have been added and all the contours have shifted; so the action of memory therefore is kind of demented and unimaginably high-speed orienteering round the landscape of the brain."[48]

As in Friel's *Molly Sweeney* and Frayn's *Copenhagen*, the problem of explaining the science directly to the audience is skillfully handled by the playwrights' sense of drama and timing: both use short, direct sentences rather than long, involved phrasing; lots of repetition; and above all many similes and metaphors that build on one another. Memory is like a map. As we get older we "lose our compass." Memory then seems like the weather; it is "unpredictable" and constantly changing. Memory becomes more an act of reassembling than retrieving stored information. In fact—and here is the dramatic climax of this part of the lecture—"remembering is essentially not only an act of retrieval but a creative thing, it happens in the moment, it's an act, an act . . . of the imagination."[49] Thus in the course of a few sentences Simon has steered the audience not only to an understanding of how memory works according to the most recent scientific findings, but to a sense of the connection between performance and memory, acting and reenactment, theater and our inner selves. This establishes the framework for what follows, when we are asked to perform certain acts in the course of this lecture, and the science lecture at the beginning of the play thus furnishes the starting point for the theatrical event as a whole.

But the main point of the play, and its soaring achievement, is that it moves from science on the microscopic level—the chemical processes in the brain that create, store, and activate memories—to the macro-level of how time works in the universe. The plays takes us on a cosmic tour from the minutiae of our own lives to that of a five-thousand-year-old body found in the Alps to the interconnectedness of all generations—all in one fluid movement. In doing so, it comes very close to achieving the state Peter Brook has so eloquently described as the ideal theatrical experience, in which audience and actors collaborate to produce "powerful emotion," which itself is productive. "The artist is not there to indict, nor to lecture, nor to harangue, and least of all to teach," writes Brook in *The Empty Space.* "When a performance is over, what remains? Fun can be forgotten, but powerful emotion also disappears and good arguments lose their thread. When emotion and argument are harnessed to a wish *from the audience* to see more clearly into itself—then something in the mind burns. The event scorches on to the memory an outline, a taste, a trace, a smell—a picture."[50]

Mnemonic appeals directly to the audience's own association with certain objects and the memories they trigger. The emphasis on objects begins when you enter the theater. Each seat in the house has taped to its back a small plastic package containing two items: a leaf and a pair of cloth eye covers like passengers used to get in airplanes to help them sleep. A message on the package instructs us not to open it until we are told to do so. In the biochemistry lecture that opens the play, we are asked to open the package, put on the eyeshades, and hold the leaf, then think back to various points in our childhood. It becomes clear that the leaf and the eyeshades are supposed to act as mnemonic devices for us, to aid in memory recall as we are gently eased backward in time to our earliest childhood memories, then further and further back to our collective past.

Sitting in the darkened theater watching a play about time and memory unfold, we are interrupted by the request to don our black eye shields and travel back in our own personal time. In this way, Complicite introduces two very different kinds of time into the play all at once: the private kind, our own "time" that we can revisit and that belongs solely to us; and the public sense of time, cultural memories of events shared by us all. The Iceman seems to represent this latter kind, yet the play also shows movingly that he had his own inner life, which is now sealed forever but which must not be forgotten as we stare at and prod his body. The play is clearly skewed toward the personal and the private sense of time and memory: the woman's search for her father; our journey to our earliest childhood memories; and the Iceman, whose excavation becomes a very public violation of the personal dignity of the man as each nation vies for ownership of the corpse, and as the corpse itself is exposed in all its naked frailty.

In the end, Alice tells Virgil on the phone that she found her father asleep under a tree in a village, but on the verge of waking him she withdrew and gave up her quest. She could not go back and recapture this part of her past, a theme that *Arcadia* also shows. Perhaps she shrinks from meeting him because her imaginary idea of him can never be reconciled with who he actually is. Or perhaps it is all a dream.

Physical and visual transformation occurs throughout the play. For example, at various points, Virgil becomes the Iceman by stripping naked and lying on a table, where he is examined by archaeologists and reporters. The many tattoo patterns discovered on the Iceman's body are projected onto Virgil/Iceman's naked back.[51] Virgil thus physically links the modern and the ancient in his transformations on stage, and in so doing telescopes vast stretches of time into seconds.

Mnemonic shows this in another way, too: the presence on stage of many different objects, each with significance to the recovery of the past. Not only does *Mnemonic* bear a similarity with *Arcadia* in its themes, which include the idea of chaos theory as a metaphor and the notion of the urecoverability of the past, but the two plays also share the temporal trick of using a simple table that becomes the site of the past and present intermingling. We have already seen how this functions in *Arcadia*. In *Mnemonic*, not objects but human bodies inhabit the table: by simply stripping naked and lying down on it, Virgil becomes the five-thousand-year-old Iceman. At one point Virgil's role as the bridge between the two stories becomes much like the final scene of *Arcadia*: he appears in a scene with characters from Alice's story who do not interact with him. In one episode, Alice meets a BBC reporter on a train, and they sleep together. As they remove their clothes, Virgil "is between them. We do not know if it is for real or in Virgil's imagination. Alice outlines the man's naked body with her hand. Virgil almost brushes them as he comes off the bed. They lie down. . . . Virgil lies on the table USC [upstage center], in despair."[52]

The table becomes a device that permeates all boundaries of time and place. Its paradoxical, incongruous status as both solid and fluid is underlined throughout and culminates in a final, breathtaking gesture of the kind of physicality that has become Theatre de Complicite's trademark: "One by one, the members of the company follow each other in lying in the place of the Iceman. They lay themselves down [on the table] and roll off again, just as generation succeeds generation in a never-ending cycle."[53] The actors take turns, fluidly getting on and off the table, and "the rolling becomes quicker and quicker" until "finally the last company member rolls over the table. The table itself hurtles offstage."[54]

For all its spectacular manipulation of time on stage and its careful setting up of the relationship between the act of memory recall and the enactment of performance, Mnemonic is ultimately concerned with the problem of the erosion of human empathy. In a very moving final image, the scientific (here synonymous with cold and relentless) scrutiny of the Iceman's naked body gives way to a compassionate empathy. The cast gathers around the naked figure on the table, and they

> hold up the metal frame of the Iceman's refrigeration unit in front of them. They are looking at him in his museum in Bolzano. Alice joins them. Through her VO [voice over] we begin to understand they are looking at him not with mere ghoulish curiosity, not in horror, but with empathy. . . . Suddenly, strangely, one of the people behind the frame slips under it and continues towards the Iceman as if drawn into his presence. In one moment they have changed places. The man watching is on the table. He has put himself in the Iceman's place and the Iceman has become naked Virgil again.[55]

Although theatrically quite different, this ending is similar in tone to that of Arcadia. In Arcadia, the final waltz of the two couples evokes a tenderness that spans both periods and all time, superseding the intellectual fireworks of the preceding scenes. This is especially poignant because we know that Thomasina will die that very night. Time is too short, emotional connection too rare. But where Arcadia stages this theme in a relatively conventional proscenium framework, Mnemonic directly engages the audience—immerses it—through the opening exercise of memory recall, which sets up an interactive framework for the rest of the performance.

Both Mnemonic and Arcadia are concerned with the idea of recovering the past. Both show that while it may be desirable and undeniably human to try to recover the past—whether one's own or someone else's—very wrong and damaging conclusions can result in such attempts at historical investigation, of time travel. For what the investigator, the historian, fails to take into account is the "noise" that may interfere with and affect the outcome of the events he or she is researching or trying to recover. Sitting in the theater, we are experiencing how this works through the very nature of theater itself. Peter Brook writes of the mutual "assistance" of the theatrical experience, when the audience is truly engaged and provides "the assistance of eyes and focus and desires and enjoyment and concentration" and thus helps the actors to create a performance. "Then the word representation no longer separates actor and audience, show and public: it envelops them: what is present for one is present for the other."[56]

Brook further articulates this unique function of theater: after a powerful performance, "it is the play's central image that remains, its silhouette, and if the elements are rightly blended this silhouette will be its meaning, this shape will be the essence of what it has to say. When years later I think of a striking theatrical experience I find a kernel engraved on my memory: two tramps under a tree, an old woman dragging a cart, a sergeant dancing, three people on a sofa in hell—or occasionally a trace deeper than any imagery. I haven't a hope of remembering the meanings precisely, but from the kernel I can reconstruct a set of meanings."[57]

The ephemerality of the stage is exactly like that of our memories; each serves as a metaphor for the other. Theater "only lives on in people's memories," explains Michael Boyd, director of the Royal Shakespeare Company; and Julian Crouch demonstrates that "the essential act of theater can never be photographed because it's invisible."[58] Grasping at our memories of the play after the event mimics the process of recalling our own past experiences—in each case, a conscious, subjective, and always shifting reconstruction.

Infinite Proof

"Mathematics provides a new language for the theater," says Luca Ronconi, director of *Infinities*. Barrow's play does for mathematics what Frayn's *Copenhagen* did for physics. Where *Copenhagen* enacts uncertainty and complementarity, *Infinities* stages some of the great paradoxes or "thought experiments" about infinity: the Hotel Infinity in all its vastness, the notion of time travel, the idea of living forever, Borges's library of Babel with its endless corridors of books. Watching *Infinities* brings such concepts to life in a stunning combination of mathematics, philosophy, science, and theatricality.

Barrow provided the text, a mixture of original passages and selections of his and other people's essays on infinity and various mathematical thought experiments. For example, he draws on writings by Nietzsche and Borges and on contemporaries like Lightman, Vilenkin, and Hawking. The acclaimed Italian director Ronconi developed the staging in conjunction with the Teatro Piccolo and Sigma Tau Foundation in Italy. *Infinities* was performed entirely in Italian at the Teatro Piccolo site specially chosen for its production, a vast, empty warehouse in the Bovisa section of Milan that formerly housed costumes and scenery for La Scala. It was also performed in Spanish in Valencia in a more conventional theater. So far, it has yet to receive its English-language premiere.

The play demonstrates the very concepts it deals with, and takes the genre of "science plays" to a new level. It presents five scenarios on different ideas of infinity, and it deliberately rejects such mainstays of traditional theater as plot and characterization. Watching it, we breathe the air of ideas and abstractions that are magically brought to life through the material possibilities of the stage—and what a stage it is. *Infinities* is deeply tied to the extraordinary space in Milan in which it was developed and performed. The audience is led through this cavernous building, which is divided into five very different stages for each of the five scenarios on infinity. We are admitted to a new scenario in small groups every fifteen minutes, making the play theoretically infinite in its structure as it lacks a beginning, middle, and end—an open-endedness that neatly captures the thematic core of the play.

While some scenarios work better than others, *Infinities* as a whole offers a new experience for actors and audience alike in which the language of drama and the language of science spectacularly meet.

Scenario 1, "Welcome to the Hotel Infinity," dramatizes a famous thought experiment by the mathematician Dave Hilbert. At Hilbert's Hotel, an overwrought manager with an infinite number of rooms must accommodate increasing numbers of guests in increasingly complex numerical arrangements, from one new guest to whole galaxies of new arrivals. The actors explain the complex mathematics with the help of a huge monitor displaying the equations necessary to work it all out. With tongue only slightly in cheek, Barrow concludes with a cosmological moral: Hotel Infinity to Hotel Zero? If the universe is infinite and began from nothing, maybe it will one day return to nothing after it gets too complicated and unmanageable to keep it running anymore.

The audience's next stop (if you are in this particular group) is about living forever. We are in a claustrophobia-inducing black box full of impossibly old people languidly reading in their wheelchairs or under salon-style hair dryers. The stifled atmosphere and long monologues create a

Infinities, scenario 1: The cavernous, soaring space in which this scenario is set helps to convey visually the thought experiment it treats, namely, the Hotel Infinity (Hilbert's Hotel). The agitated hotel manager must figure out a way to accommodate an infinite number of guests. Photo © Serafino Amato.

monotony that effectively conveys the idea of perpetuity, and how unappealing it would be to go on living forever. Living infinitely would also raise some interesting questions. How would pensions be managed? Would we need life insurance, and if so, how would the cost of it be worked out?

Scenario 3 dramatizes Jorge Luis Borges's parable of the Library of Babel. Visually inventive staging, involving the use of mirrors at the ends of several corridors lined with drawers large enough to hold a human body, creates the illusion of an infinite library, and the audience is invited to wander from corridor to corridor while the voices of the actors resound

Infinities, scenario 3: Borges's Library of Babel provides the textual impetus for exploring the notion of infinite knowledge as well as the Duplication Paradox thought experiment, here enacted using scenography as well as half masking, conveying the idea of identical doubles. Photo © Serafino Amato.

around them. Another neat trick is the appearance of an increasing num-ber of identically masked and clothed actors—a disturbing effect as the text addresses the Duplication Paradox suggesting the impossibility of uniqueness and individuality. The actors underscore the message of the text by endlessly replicating themselves.

The fourth scenario stages the famous conflict between the nineteenth-century mathematicians Cantor and Kronecker. The textual narrative of Cantor's troubled life is illuminated visually by a wheelchair-bound actor swathed in white bandages who sits immobile under the rantings of his

Infinities, scenario 5: In this scenario on time travel, a long line of books in crates on the floor symbolizes time and history across which the actors run and leap, while an actor demonstrates the Grandmother Paradox—why you cannot go back in time and kill your own grandmother (or yourself).
Photo © Serafino Amato.

assailant, Kronecker—all of this taking place on tables in a simulated classroom, with the audience seated at these tables among real students from the local polytechnic who rise now and then to paint the mathematical equations and names of the great mathematicians the text refers to on huge white sheets of paper on the walls.

Just when you think the play cannot possibly get any more inventive, you enter the gigantic open space of the final scenario, which is about time travel. A grandmother teeters across the stage, at one point narrowly missing the grandson speeding toward her in his wheelchair (illustrating the Grandmother Paradox). An actor leaps effortlessly through time, which is symbolized by a long span of book-filled crates. The concept of time traveling is further literalized through the use of a train car complete with preset pop-up dining tables at which passengers sit facing in both directions.

The theatrical implications of Barrow's and Ronconi's play will be discussed further in the concluding chapter, which also focuses on the techniques of Theatre de Complicite and other companies that are taking science on stage in new directions. It remains to be said that the three widely diverse and theatrically striking and original plays discussed here demonstrate that science plays that engage mathematics may well provide some of the most interesting and innovative examples of the genre we have yet seen.

7 Doctors' Dilemmas: Medicine under the Scalpel

●●●●●●●●●●●●●●●●●●●●●●

A chapter on plays dealing with medicine and doctors clearly belongs in this book for two main reasons: first, this branch of science plays offers a wonderfully diverse array of engaging and provocative dramas; second, such plays have a special relevance for many audience members for whom doctor-patient interaction is often their only direct, personal encounter with science. We have all been to the doctor; few of us have visited a particle accelerator or know what a quark is. A substantial portion of science plays deals specifically with the medical profession and explores doctor-patient relationships on stage. The more recent of these include Margaret Edson's *Wit*, Brian Friel's *Molly Sweeney*, and Peter Brook and Marie-Hélène Estienne's *The Man Who*. Others, like Tony Kushner's *Angels in America*, include much detailed medical knowledge while foregrounding other, more political, themes. All these more or less fictionalized medical or "doctor" plays build on a tradition of depicting the medical profession on stage.[1] Molière, for example, "seems to have been fascinated" by doctors, depicting "a therapeutic world in which the real doctors fail, the fake doctors succeed in healing, and the difference between doctor and patient is not always clear."[2] Shaw's *The Doctor's Dilemma* is perhaps as indebted to Molière's paradoxical medical world as it is to Jonson's *The Alchemist*. As we shall see later, there is also a variety of doctors and diseases in Ibsen's plays, from *A Doll's House* to *Ghosts* to *An Enemy of the People*.

The idea of staging doctors in a more realistic way, showing what they actually do in their interactions with patients, comes to the fore in the early part of the twentieth century, alongside the rapid rise of modern medicine. This is reflected in plays like Eugene Brieux's *Les Avariés* (1909), Sidney Kingsley's *Men in White* (1934), and the Federal Theater Project's *Medicine Show* (1940). This latter work was produced for Broadway, and in it "the protagonist, Mr. Average Citizen, toured a modern hospital. Throughout, both he and his guide remained stationary. As each new section of the hospital was presented, appropriate pictures of instruments, apparatus, and patients were projected. The result was to empha-size the hospital environment by showing it in larger-than-life detail."[3] Modern medical dramas tend to appeal to the audience's apparent curios-ity about hospitals and the medical profession behind the scenes. A newer development is the group of plays built on interviews with patients, such as Nell Dunn's *Cancer Tales*—"true stories told in our own words"[4]—and Anna Deavere Smith's untitled one-woman show about the experiences of physicians and their patients at a big urban hospital.[5]

It is striking that of all the plays about doctors and patients mentioned here, only Kingsley's offers an unreservedly positive portrayal of the medi-cal profession. Even Anna Deavere Smith's untitled play, in which she performed extracts from her interviews with doctors and patients at Yale–New Haven Hospital to convey both perspectives on these relationships, seemed to portray doctors less sympathetically than patients. The text came verbatim from interviews with doctor and their patients—nothing that was said was made up by the playwright—yet for all its objectivity the show revealed an uncomfortable imbalance of power in favor of the doctors. It is hardly surprising, then, to find a recent play, *The Seventh Chair*, imagining Ibsen's Dr. Stockmann and several other well-known dramatic doctors attending group therapy "for doctors who are experienc-ing a crisis in professional identity."[6]

Why do doctor-patient plays generally come down hard on doctors, who after all have gone through years of grueling training expressly to help people and save lives? Certainly Hollywood is partly responsible; as James Welsh points out, films tend to draw on the same tired stereotypes of the greedy, self-absorbed doctor or the medical madman.[7] Perhaps such negative portrayals of doctors are an attempt to capture and redress an implicit imbalance of power, in which patients are at the mercy of doctors and their knowledge, and have to depend on and trust in the doctor's will to use that knowledge for the patient's good. Whatever the reasons, this chapter will explore how the medical-playwriting trend relates to science and theater generally at a time when plays about scientific subjects have proliferated and are beginning to receive critical attention.

Ibsen's Impotent Doctors

Ibsen's *An Enemy of the People* (1882) features Stockmann, a doctor-scien-tist who discovers that his town's spa baths—its livelihood—are full of "putrefying organic matter" and "millions of infusoria."[8] If they are not closed down and treated, people will surely die, yet the townspeople are outraged at having to take such costly measures. Far from being hailed as its savior, Dr. Stockmann is castigated by the town for threatening its livelihood through his scientific discovery. His own brother, the Mayor, joins forces with the townspeople and the so-called liberal press to pillory him. Doctors usually come under criticism in Ibsen's plays, as in many of Chekhov's, and Stockmann is no exception. Like Ibsen's other doctors he is ineffectual and impotent. He detects the ills of society but can do nothing to remedy them.

The societal status usually accorded doctors proves worthless in Ibsen's plays; it is undermined by the emptiness of such alleged authority, since they can do nothing. While many readers and audiences see Dr. Stockmann as a hero, an interpretation Ibsen promoted by the pointedly auto-biographical elements in the character, Stockmann is ambivalent at best. Initially, he is attractively portrayed—youthful and energetic; infectiously enthusiastic about science; good at it (he is careful enough to verify his findings about the waters through an outside lab before releasing them to the public); a lover of life, good food, and drink; a family man; and encouraging of the young and their new ideas. Then it becomes clear that his motives are not quite disinterested; he is full of what theater scholar Thomas Van Laan has identified as "problematic qualities."[9] Vain and self-righteous as well as naive, Stockmann professes to some disturbingly aristocratic views about the masses and the common people as "vermin" and "curs" that border on the rhetoric of eugenics. He shudders with hatred when he recalls his exile as a doctor in the remote northern parts of Norway, tending to the "vermin." As Van Laan points out, his uncomfortable predecessor is Shakespeare's Coriolanus, who is repeatedly called "enemy of the people" and also regards the common people as "curs." Several times Stockmann drops pointed hints about how the town will inevitably show its gratitude for his discovery about the baths—"a demonstration in my honour, or a banquet, or a subscription list for some presentation to me"—yet feigns indifference to such tributes.[10] This difficult doctor, torn between a desire to help people and an aristocratic disdain for them, encompasses both the best and the worst of his profession.

In general, Ibsen's doctors are expert diagnosticians but inept curers. The dying Dr. Rank in A Doll's House and the drunken Dr. Relling in The Wild Duck bear this out. Both function as family friends and raisonneurs in the respective plays. Dr. Rank succumbs to an inherited disease he is able to diagnose but powerless to stop, which serves as a metaphor for his role in the Helmer household, wisely observing the foibles of the various family members but unable to hinder their mistakes. Dr. Relling is clever at diagnosing the problem with all of humanity—our need for self-deception, the "life lie"—but stumbles about drunkenly, unable to do anything to stop the disaster he all too well sees coming in the Ekdal house. Awareness coupled with passivity makes these doctors doubly culpable, Ibsen seems to be saying, as opposed to characters who lack such sharp insights into their own weaknesses and motives. Thus, in the complex moral universe Ibsen devises, we can perhaps find greater capacity to empathize with the deeply flawed Hjalmar Ekdal, whose ignorance and self-deception lead to Hedwig's suicide, than with the insightful Dr. Relling, who can see what is wrong with people and with society but has no power to change or heal them.[11]

Syphilis has featured in several plays over the last century or so, beginning with Ibsen's *Ghosts* (1881), in which Oswald, the young artist son of the unhappily married and now widowed Mrs. Alving, is afflicted with the terrible disease. It is clearly depicted as a hereditary illness; the idea of the "sins of the father" being visited on the son dominates this play, reflecting the received opinions about syphilis at this time. Oswald tells his mother about his illness and shows her the fatal dose of morphine he has managed to procure and is requesting her to administer to him as soon as he shows signs of "brain softening"; he wants her to kill him mercifully rather than allow him to succumb to the disease. The play thus also introduces euthenasia as a dramatic theme as well as sexually transmitted disease. The final moments of the play are harrowing: the curtain descends as Mrs. Alving, holding the vial in her trembling hand, stares in horror at Oswald, who has slumped in his chair and is tonelessly gibbering "the sun, the sun" over and over. Will she or won't she? The audience is left having to guess, and both options are appalling and tragic.

Ghosts is one of several of Ibsen's plays that depicts the effects of syphilis. Its predecessor, *A Doll's House* (1879), includes a subplot featuring a doctor dying of the disease, which he has diagnosed himself and suspects was hereditary. "And to suffer thus for another's sin! . . . My poor innocent spine must do penance for my father's wild oats," says Doctor Rank.[12] In *The Wild Duck* (1884), Ibsen alludes to syphilis having caused the encroaching blindness of Hedwig, the thirteen-year-old illegitimate daughter of a philandering aristocrat and the serving girl he seduced. Syphilis furnishes Ibsen a powerful metaphor for the twin ideas of inherited sins and genetic illness corrupting the innocent.

One fact about STD plays that is immediately striking, though hardly surprising, is how many people they have reached. But it is not just because they were notorious for their subject matter or their frank approach to it; it also has to do with the kinds of theaters that put them on and the fact that they have also had an avid readership. Brieux's *Les Avariés* (translated variously as *Shipwrecked* or as *Damaged Goods*) was performed in Germany between 1910 and 1920 and reached a total audience of one million in that country alone.[13] The play *Spirochete* (published in 1938), which traces the origins and development of syphilis, reached an audience of thousands as one of the Living Newspapers produced by the Federal Theater Project. As noted in chapter 3, the FTP was able to reach millions of people through its productions in many of America's largest cities. *Spirochete* incorporates excerpts from another play into its text—in this case, key scenes from *Les Avariés*. Thus Brieux's play was given added life through its featuring in the FTP's production about syphilis.

Although frequently compared with *Ghosts*, Brieux's syphilis play, *Les Avariés*, pales beside Ibsen's. Brieux goes into much greater scientific detail about the disease and conveys much more information for the audience, but *Les Avariés* is theatrically leaden, containing long passages of explanation unleavened by wit or dramatic effect. It is wholly a thesis play, with characters whose lines read as if straight out of medical textbooks of the time. The play is thoroughly conventional in its style; that and not the topicality of its subject is the real problem. After all, Ibsen's *Ghosts*, also "about" syphilis (in part), has not lapsed out of date at all, indeed has taken on renewed relevance with the onset of AIDS, whereas *Les Avariés* has gone stale and wooden. While Shaw was right in pinpointing the scientific climate of the age, the passage of time has surely proved him wrong in his extravagant assessment of Brieux as "the most important dramatist west of Russia" after Ibsen's death, and "incomparably the greatest writer France has produced since Molière."[14] Indeed, time has shown Brieux to be precisely the "mere pamphleteer without even literary style" that Shaw tried to defend.[15] It is no surprise that Ibsen's plays, despite some rather confused notions about syphilis, make far better drama and have survived the years much better than Brieux's.

Despite the passage of time, sexually transmitted disease is still controversial in the theater. Tony Kushner's two-part play *Angels in America* was a huge hit on Broadway and on tours across America, not just on its own merits but also because of the adverse publicity its explicit portrayal of AIDS and homosexuality generated, especially in more conservative areas of the country. The play was almost immediately anthologized in textbooks for college drama and theater courses, thereby reaching a huge readership in addition to its audiences; now the 2003 TV film version (seven hours long) starring Meryl Streep, Al Pacino, Emma Thompson, and Mary Louise Parker has given the play an even higher profile in a rare example of a successful crossover from play to screen adaptation (*Wit* is another, more modest, example). The public demonstrations against various productions of *Angels in America* testify to the power of a play that broke new ground in opening up the discourse on sexually transmitted diseases through in-depth dialogue about AIDS and HIV, and not just set in hospitals or doctors' examining rooms but among lovers, families, friends, and colleagues in a variety of settings. It is a great example of how STD plays work theatrically and how they have entered and contributed to the discourse on this sensitive subject.

Stimulate the Phagocytes!

Shaw may have been carried away in his assessment of Brieux, but he is a key figure in the development of "medical plays" through his own *The Doctor's Dilemma* and his theory of creative evolution and the "life force" as shown in *Man and Superman* and *Candida*. In *The Doctor's Dilemma* (1906), a triumphant updating of Jonson's *The Alchemist*, Shaw created a memorable group of physicians, most of them scurrilous and out to dupe gullible patients for their money. The play is a prime example of medical dramas that concern themselves centrally with how physicians treat patients, and its hostile attitude toward doctors makes its polemical target clear from the start. Two patients need a cure for tuberculosis: Louis Dubedat, a gifted but (by conventional standards) amoral artist, and Blenkinsop, a colorless and plodding country doctor. Dubedat is charming but feckless, a liar, a cheat, and a polygamist, but he creates stunningly beautiful art; Blenkinsop is drab and humorless, professionally unfashionable but dutifully healing the sick and saving lives. The "dilemma" is which one of these should receive the precious experimental treatment pioneered by Ridgeon, a genial and likable doctor who has just been knighted as the play opens.

Ridgeon has been knighted for a "great discovery" in the field of biochemistry, relating to phagocytosis. "The phagocytes won't eat the [disease] microbes unless the microbes are nicely buttered for them," he explains to his old friend Sir Patrick, the sensible, old-fashioned doctor who functions as *raisonneur* and also aids in the play's exposition. "Well, the patient manufactures the butter for himself all right; but my discovery is that the manufacture of that butter, which I call opsonin, goes on in the system by ups and downs . . . and that what the inoculation does is to stimulate the ups and downs, as the case may be." The upgrade is positive, the downgrade is negative, and "everything depends on your inoculating at the right moment. Inoculate when the patient is in the negative phase and you kill: inoculate when the patient is in the positive phase and you cure."[16] You test the patient's blood to get his or her opsonin index. "If the figure is one, inoculate and cure: if it's under point eight, inoculate and kill. Thats my discovery: the most important that has been made since Harvey discovered the circulation of the blood. My tuberculosis patients dont die now."[17]

The dilemma is complicated by the unpredictable Russian-roulette aspect of the cure: it works only if the patient's body is in the right state, and that cannot be predicted. If the body is not in the right state, the treatment will not work, and the patient will die. Ridgeon sums up his predicament: "It's not an easy case to judge, is it? Blenkinsop's an honest decent man; but is he any use? Dubedat's a rotten blackguard; but he's a genuine source of pretty and pleasant and good things. . . . It would be simpler if Blenkinsop could paint Dubedat's pictures."[18] In love with Jennifer, the beautiful young wife of Dubedat, and offended by the husband's moral fecklessness, Ridgeon decides to administer his treatment to Blenkinsop and to leave Dubedat in the hands of a colleague. Dubedat dies, and in the final scene Ridgeon goes to woo the young widow at a posthumous exhibit of Dubedat's art that she has mounted, only to be firmly rebuffed. Shaw poses, but never fully solves, the riddle of why Jennifer Dubedat, clearly aware of her husband's failings and his deception of her, not only tolerates but loves and defiantly champions him. She seems to be Shaw's strongest voice for Art, an Art that is apart from morality. The point is not that Art triumphs over Science, but that Shaw pits the two against one another in the first place, and stacks the cards so totally in favor of Art.

Why does he do this? Is it to expose the hypocrisy and uselessness of the conventional moral values that are self-righteously espoused by the group of doctors and willfully flouted by the young artist? The play's weak moments, such as a thinly veiled lecture on antivivisectionism and Shaw's jarringly intrusive self-reference in the middle of the play, are offset by its engagement with such serious philosophical issues. *The Doctor's Dilemma* poses an impossible moral problem, asking how we can make a moral choice in an unethical situation, a theme of almost all science plays, and it brings it up to date with its own audience by utilizing the most contemporary scientific references and recent discoveries. Although Shaw's moralizing can be heavy-handed, he includes suggestive touches such as the "curious aching" that Ridgeon suffers from the outset of the play, auguring tragedy and loss.[19]

However personable and attractive he may seem, Ridgeon in his new-found glory has strayed from the Hippocratic oath. Shaw's punishment of doctors, his vehement "hatred of the medical profession and scientific medicine," ought to be placed in context with his own experience as a smallpox sufferer inadequately protected by the vaccine that was supposed to prevent the disease, leaving "psychological scars" that found expression in such writings as the hundred-page preface to *The Doctor's Dilemma* and a later collection of articles called *Doctors' Delusions.*[20] "[He] specifically attacked Edward Jenner, Louis Pasteur, and Joseph Lister," whose fame "rested on controlling micro-organisms."[21] All this invective against doctors and scientists came from one general source: "a peculiar sense of being assailed by an unseen world of germs, which he evidenced in a virulent hypochondria."[22] But Shaw got to know his enemy by teaching himself some basic biochemistry, or what would have been known at the time about microbes.[23] This research finds expression as his characters exclaim things like "Stimulate the phagocytes!" and casually discuss tuberculosis, hematology, and opsonin as part of the play's exposition:

> SIR PATRICK: Opsonin? What the devil is opsonin?
>
> RIDGEON: Opsonin is what you butter the disease germs with to make your white blood corpuscles eat them.
>
> SIR PATRICK: That's not new. I've heard this notion that the white blood corpuscles—what is it that what's his name?—Metchnikoff?—calls them?
>
> RIDGEON: Phagocytes.
>
> SIR PATRICK: Aye, phagocytes; yes, yes, yes. Well, I've heard this theory that the phagocytes eat up the disease germs years ago.[24]

Joined by another colleague, Sir Ralph Bloomfield Bonington ("B.B." to his friends), the doctors continue to discuss medical findings and treatments: "Just as men imitate each other, germs imitate each other. There is the genuine diphtheria bacillus discovered by Lœffler; and there is the pseudo-bacillus, exactly like it, which you could find, as you say, in my own throat." The way you distinguish the two is simple: "If the bacillus is the genuine Lœffler, you have diphtheria; and if it's the pseudo-bacillus, youre quite well."[25]

Just as there are genuine and pseudo-germs, so too are some doctors real and some fake. Shaw follows this serious exchange between scientists and doctors with dialogue spoken by Walpole, a successful surgeon and a true charlatan reminiscent of *The Alchemist*'s Subtle. Already Walpole's sartorial choices are at odds with the conventional, sober image of the medical doctor; he is "smartly dressed with a fancy waistcoat, a richly colored scarf secured by a handsome ring, ornaments on his watch chain, spats on his shoes, and a general air of the well-to-do sportsman about him."[26] His medical and financial success derive solely from his theory about blood poisoning and his discovery of the nuciform sac, a mythical organ rather like the appendix, which Walpole removes from a steady stream of patients clamoring for the treatment:

> WALPOLE: I know what's the matter with you. I can see it in your complex-
> ion. I can feel it in the grip of your hand.
> RIDGEON: What is it?
> WALPOLE: Blood-poisoning.
> RIDGEON: Blood-poisoning! Impossible.
> WALPOLE: I tell you, blood-poisoning. Ninety-five per cent of the human
> race suffer from chronic blood-poisoning, and die of it. It's as simple as
> A.B.C. Your nuciform sac is full of decaying matter—undigested food
> and waste products—rank ptomaines. Now you take my advice,
> Ridgeon. Let me cut it out for you. You'll be another man afterwards.
> SIR PATRICK: Don't you like him as he is?
> WALPOLE: No I don't. I don't like any man who hasn't a healthy circulation.
> I tell you this: in an intelligently governed country people wouldn't be
> allowed to go about with nuciform sacs, making themselves centres of
> infection. The operation ought to be compulsory; it's ten times more
> important than vaccination.[27]

Despite such deliberate silliness, spoofing doctors and the medical profession, there are serious scientific moments in the play, and it is clear that although he launches a ferocious attack on the medical establishment, Shaw has genuinely investigated the biochemistry discussed by the doctors—not just to make them convincing or because it is inherently dramatic and interesting, but to strengthen his polemical case against research scientists and doctors.

Indeed, there is a distinction in the play between scientific research and its clinical applications. Shaw seems to indicate that science itself is not the problem, but human agency. Although Ridgeon seems sincerely dedicated to developing his biochemical research into a cure for tuberculosis, in the end he too reveals his shallowness and immorality in coolly letting a patient die whose wife he is in love with in the hope that the way will be cleared for him to win her. The only comparatively positive portraits of doctors in the play are the old-fashioned and now retired doctor, Sir Patrick, whose role is to debunk the myths of the greatness of modern medicine that the younger doctors go around spreading; the illustrious B.B., into whose hands Dubedat is finally placed (and whose treatment kills him); and Blenkinsop, who by the London practitioners' superficial standards is unglamorous and unsuccessful. Yet even the affable B.B. has a disturbing attitude toward his patients:

> You ask me to go into the question of whether my patients are of any use either to themselves or anyone else. Well, if you apply any scientific test known to me, you . . . will be driven to the conclusion that the majority of them would be, as my friend Mr. J. M. Barrie has tersely phrased it, better dead. Better dead. . . . [W]hat are many of my patients? Vicious and ignorant young men without a talent for anything. If I were to stop to argue about their merits I should have to give up three-quarters of my practice.[28]

Echoing Dr. Stockmann in *An Enemy of the People*, who characterizes his patients in the North as "vermin" to be extinguished, this passage and others in *The Doctor's Dilemma* convey a troubling ambivalence about the medical profession: in theory a noble calling to heal the sick, yet in practice an often thankless, repulsive task. Yet such antiheroic depiction of doctors is thoroughly Brechtian; as his Galileo asks, why do we need heroes? Here is another instance of science plays adopting such defamiliarizing techniques.

Punctuation Matters

"Far and away the most celebrated new play of 1998,"[29] *Wit* intertwines medicine and metaphysical poetry in the depiction of a middle-aged English professor, Vivian Bearing, dying of advanced ovarian cancer. As she undergoes the severe treatment and the drastic physical and emotional changes it brings, she reflects on her life and comes to acknowledge her failure to forge emotional bonds with others. The play shows doctors treating this patient in the hospital, in a series of specific clinical interactions that provide the setting for each scene—a pelvic examination, a grand rounds session, postexamination discussions, an attempted resuscitation complete with CPR and defibrillators. The play does not just refer to Vivian's hospitalization and her symptoms; it shows them, from her hair loss to her vomiting to her becoming comatose and finally dying.

Nancy Franklin called the play "thrilling" to see; "the evening has a sub-
lime indivisibility and power that no account of particulars can, finally,
get at." It makes you think about no less than "the essential tragedy and
comedy of being human."[30]

This is a major achievement for any play, so how does the playwright
accomplish this in a work about death, cancer, and hospitals? A key theme
in the play is the relationship between life and death, but it is not referred
to in oblique, amorphous ways; Edson uses a highly specific method of
addressing it, namely, by analogy with punctuation in poetry. She brings
literary criticism to bear on science and medicine, as close textual analysis
parallels the medical investigation of the body. Vivian anatomizes poetry
as she in turn is anatomized by her doctors. The striking thing about this
is that it is an inversion of the usual method of science plays, which is to
use scientific themes to explore human questions; here, literature is used
to explore science, and poetry provides the metaphor while science and
medicine are the context for its demonstration.

Edson's specific metaphor is the contested punctuation in the final line
of a famous Donne sonnet and how it relates to Vivian's own approaching
death. "And death shall be no more, *comma*, Death thou shalt die. . . .
Nothing but a breath—a comma—separates life from life everlasting,"
Vivian's old professor, E. M., tells her in a flashback to her student days,
which sets up the metaphor and its centrality to the play as a whole.
"This way, the *uncompromising* way, one learns something from this poem,
wouldn't you say? Life, death. Soul, God. Past, present. Not insuperable
barriers, not semi-colons, just a comma." The tutorial instructs the audience as to the play's central ideas and turns several of them against one
another, complicating matters for us just as Vivian's youthful assumptions
are challenged:

> E.M.: Please sit down. Your essay on Holy Sonnet Six, Miss Bearing, is a
>   melodrama, with a veneer of scholarship unworthy of you—to say noth-
>   ing of Donne. Do it again.
> VIVIAN: I, ah . . .
> E.M.: You must begin with a text, Miss Bearing, not with a feeling.
>
> Death be not proud, though some have called thee
> Mighty and dreadfull, for, thou art not soe.
>
> You have entirely missed the point of the poem, because, I must tell you,
>   you have used an edition of the text that is inauthentically punctuated.
>   In the Gardner edition—

VIVIAN: That edition was checked out of the library—

E.M.: Miss Bearing!

VIVIAN: Sorry.

E.M.: You take this too lightly, Miss Bearing. This is Metaphysical Poetry, not The Modern Novel. The standards of scholarship and critical reading which one would apply to any other text are simply insufficient. The effort must be total for the results to be meaningful. Do you think the punctuation of the last line of this sonnet is merely an insignificant detail?

The sonnet begins with a valiant struggle with death, calling on all the forces of intellect and drama to vanquish the enemy. But it is ultimately about overcoming the seemingly insuperable barriers separating life, death, and eternal life.

In the edition you chose, this profoundly simple meaning is sacrificed to hysterical punctuation:

And Death—*capital D*—shall be no more—*semi-colon!*
Death—*capital D*—comma—thou shalt die—*exclamation point!*

If you go in for this sort of thing, I suggest you take up Shakespeare.

Gardner's edition of the Holy Sonnets returns to the Westmoreland manuscript source of 1610—not for sentimental reasons, I assure you, but because Helen Gardner is a *scholar*. It reads:

And death shall be no more, *comma*, Death thou shalt die.

(As she recites this line, she makes a little gesture at the comma.)

Nothing but a breath—a comma—separates life from life everlasting. It is very simple really. With the original punctuation restored, death is no longer something to act out on a stage, with exclamation points. It's a comma, a pause.

This way, the *uncompromising* way, one learns something from this poem, wouldn't you say? Life, death. Soul, God. Past, present. Not insuperable barriers, not semicolons, just a comma.

VIVIAN: Life, death . . . I see. (*Standing*) It's a metaphysical conceit. It's wit! I'll go back to the library and rewrite the paper—

E.M.: (*Standing, emphatically*) It is *not wit*, Miss Bearing. It is truth. (*Walking around the desk to her*) The paper's not the point.

VIVIAN: It isn't?

E.M.: (*Tenderly*) Vivian. You're a bright young woman. Use your intelligence. Don't go back to the library. Go out. Enjoy yourself with your friends. Hmm?

(Vivian *walks away*. E.M. *slides off*.)

VIVIAN: (*As she gradually returns to the hospital*) I, ah, went outside. The
    sun was very bright. I, ah, walked around, past the . . . There were stu-
    dents on the lawn, talking about nothing, laughing. The insuperable
    barrier between one thing and another is . . . just a comma? Simple
    human truth, uncompromising scholarly standards? They're *connected*?
    I just couldn't . . .
    I went back to the library.[31]

Vivian is subjected to a rigorous criticism of her own shoddy scholarship
and then, poised to rewrite the paper with the proper standards, unexpectedly and very gently told that there are more important things to attend
to: socializing and connecting with others. Donne alone cannot nurture
her soul. This sets up an opposition between academic pursuit and "normal" human existence, as if the two are incompatible—as if to show that
in chasing her academic dream Vivian misses this crucial point about
connectedness to her fellow human beings.

But it would be too simplistic to read this scene and the play as a
whole as a criticism of academics. Edson's heroine may be cold, tough,
and uncompromising, but it is not because she is an academic; it is not
because she engages in intellectual pursuit, but because she has misunderstood its purpose. Vivian has essentially used her devotion to the highest
intellectual standards as an excuse to shun or demean all those who fall
short of them, whether students or colleagues; but the target here is not
intellectuals, it is anyone who is so exacting and rigid. This is no trite
theme for a contemporary American play, it is a valuable and timely message to send out in an increasingly anti-intellectual climate. Far from
condemning intellectuals, Edson upholds the pursuit of learning and asks
simply that humanness not be sacrificed on the altar of scholarship or
indeed any endeavor. "Now, to the doctors," writes one critic, "Vivian is
the recipient of such dismissiveness [as she used to show toward her stu-
dents]; *she* is the text: something to be examined, studied, and experi-
mented with."[32]

In fact, the play displays the need for literature and art to balance the
more practical aspects of life, just as Shaw's *The Doctor's Dilemma*
teaches (albeit far more polemically) the lesson that art fulfills a vital
and unappreciated role in enriching life, particularly the moral life, of
individuals and of society as a whole. That Vivian can learn such a lesson
so late in life, can change for the better and rectify her wrongs in rebuffing
people's advances, speaks for this interpretation of the play. Far from remaining rigid and disconnected when facing her ordeal with cancer, she

is able to grow and change, eventually fulfilling what E.M. had so tenderly asked her to do: go out and make some friends, enjoy the sunshine, be human. She can only do this within the confines of the cancer ward, but make friends she does, most poignantly with the decidedly unliterary Susie, the primary nurse. Susie knows Vivian will die soon, but she still befriends her, brings her comfort, and eases her pain through small but tender gestures such as massaging cream into her hands.

The very fact that the distinguished professor E.M. conveys this important life lesson is a rebuttal of any contention that academics are the target of criticism in this play. In fact, Vivian's two doctors come under the harshest criticism for precisely the same failing as Vivian's—they are so preoccupied with their experiment on her, with achieving success for themselves through her body, that they neglect her feelings and come across as cold, tough, and inhuman. Like Vivian, her doctors, both senior and junior, are highly successful and driven people of great intelligence, yet they are finally cold and uncaring, as she has been until now. Dr. Kelekian is "much more interested in research than he is in people," while his protégé, Dr. Jason Posner—a former student of Vivian's—"actually has to remind himself to ask his patient how she is feeling . . . [and] regards Bearing more or less as packaging—it's her cells he cares about."[33] Jason took Vivian's course because he knew that it would look good on his curriculum vitae if he could get an A from one of the university's most demanding teachers (he got an A−), thus helping him get into medical school. One of the most humbling experiences Vivian has is seeing her own youthful misunderstanding of E.M.'s toughness relived in Jason's emulation of *her* tough and uncompromising attitude toward other people.

A related opposition is set up in this early scene (excerpted earlier) between wit and truth. Edson seems to be drawing attention to the emptiness and contrived quality of pure wit and indicating that Vivian's admiration of it is misplaced and has led to her prizing "metaphysical conceits" over emotional depth. As in so many science plays, there is a metatheatricality at work here. Edson calls attention to the artifice of theater, the heightened emotion of melodrama, as something devoid of truth and true emotion, to be eschewed in favor of calm, introspective self-awareness and reflection. The very structure of the play reinforces this, as Edson forgoes straight, lulling realism in favor of inventive ways of breaking the fourth wall: in addition to speaking directly to the audience and using asides in dialogues with other characters, Vivian moves into flashbacks wearing her hospital gown, which creates a discordant, incongruous note and reminds the audience constantly of the artifice of the stage.

The witty dig at Shakespeare even has a function, for while in this scene E.M. dismisses him as melodramatic and histrionic, her last words in the play—her adieu to Vivian—are Horatio's final words to the dead Hamlet: "And flights of angels sing thee to thy rest." This line is immediately followed by the traumatic final scene of *Wit*, in which the highly dramatic, frenzied actions of the doctors working to revive Vivian are set against the calm, serene movements of the character as she walks slowly downstage to meet the life everlasting. The references to Shakespeare and to acting and the stage in general in the earlier scene quoted previously have helped to lay a foundation for the enactment of the metaphor at the end of the play. In these final moments, Edson makes an elegant thematic connection between form and content by having the climactic final scene of a mistaken attempt to revive Vivian on her deathbed literally perform the punctuation discussed in analyzing the sonnet.

The final scene begins when Jason has come into Vivian's room for a routine check and realized that she shows no vital signs. His actions become violent: he "throws down the chart, dives over the bed, and lies on top of her body" to reach the phone to call a code blue team to revive her. Then he begins CPR on her, "pounding frantically" on her chest. Susie enters, and a fierce, violent struggle ensues. Jason is pushed off Vivian's bed onto the floor while the code team mistakenly goes to work on Vivian and Jason realizes his error:

JASON: (*In agony*) Oh, God.
CODE TEAM:—Get out of the way!
    —Unit staff out!
    —Get the board!
    —Over here!
    (*They throw Vivian's body up at the waist and stick a board underneath for CPR. In a whirlwind of sterile packaging and barked commands, one team member attaches a respirator, one begins CPR, and one prepares the defibrillator. Susie and Jason try to stop them but are pushed away. The loudspeaker in the hall announces "Cancel code, room 707. Cancel code, room 707."*)
CODE TEAM: —Bicarb amp!
    —I got it! (*To Susie*) Get out!
    —One, two, three, four, five!
    —Get ready to shock! (*To Jason*) Move it!
SUSIE: (*Running to each person, yelling*) STOP! Patient is DNR!
JASON: (*At the same time, to the Code Team*) No, no! Stop doing this. STOP!

CODE TEAM:—Keep it going!
  —What do you get?
  —Bicarb amp!
  —No pulse!
SUSIE: She's NO CODE! Order was given—(*She dives for the chart and holds it up as she cries out*) Look! Look at this! DO NOT RESUSCITATE. KELEKIAN.
CODE TEAM: (*As they administer electric shock,* Vivian's *body arches and bounces back down.*)
  —Almost ready!
  —Hit her!
  —CLEAR!
  —Pulse? Pulse?
JASON: (*Howling*) I MADE A MISTAKE!
  (*Pause. The* Code Team *looks at him. He collapses on the floor.*)
SUSIE: No code! Patient is no code.
CODE TEAM HEAD: Who the hell are you?
SUSIE: Sue Monahan, primary nurse.
CODE TEAM HEAD: Let me see the goddamn chart. CHART!
CODE TEAM: (*Slowing down*)
  —What's going on?
  —Should we stop?
  —What's it say?
SUSIE: (*Pushing them away from the bed*) Patient is no code. Get away from her![34]

The violent attempts at resuscitation amount to a battle over Vivian's body in a remarkable climactic scene in which we see her leave the bed where her lifeless body is being worked on by the code team, come downstage, undress, and reach up toward the light, fully naked for just an instant before the final blackout. While her soul serenely detaches itself, Jason and the code team have been trying to enforce a semicolon—or, worse still, an exclamation point!—where there should be only a comma; they have tried artificially to prolong the space between life and life everlasting. With the doctors flailing around her body, shouting profanities and administering wrenching shocks, she bears out the truth of what her professor had said: there should be no dramatic exclamation points at the end, only a quiet surrender, a breath between life and death.

As with plays like *Copenhagen, After Darwin,* and *Arcadia, Wit* provides another example of a contemporary science play whose dramaturgical strategy is to perform its central theme, using structural and staging devices to convey ideas.[35] The form of the play is designed to enact the theme; the two are mutually dependent. In this case, however, there is an inversion: instead of using some scientific idea as a metaphor for a universal truth about the human condition, the play uses poetry and sets it within a specific scientific and medical framework—the hospital. It is literature, not science, that forms the metaphor here.

There is another key difference from science plays like *Arcadia* and *Copenhagen*. In those plays, knowledge is a good thing, generally; its acquisition is to be encouraged, ideas are the stuff of life and passion, and it is part of our "long march" toward enlightenment. Part of what makes characters like *Arcadia*'s Hannah, Septimus, and especially Thomasina so attractive and sympathetic is their enthusiasm for learning and their acute intelligence. Yet in *Wit*, we see the negative effects of too much knowledge. Dr. Bearing is a highly successful academic whose knowledge of Donne's poems and of seventeenth-century metaphysical poetry is of the highest order; she is "one of those larger-than-life professors who are terrifyingly smart, and proud of it."[36] Yet she is a barren person in several ways, without friends or family; even her colleagues, she assumes, will relish her passing.

Edson seems to be warning of the dangers of prizing knowledge over basic human kindness, such as the nurse Susie shows toward her patients. Apart from E.M., Susie is the only caring person on the stage, bringing relief to the sick and above all showing compassion and concern. In the end Vivian is tended compassionately by Susie and her former tutor, E.M., her only visitor, who offers the greatest comfort to Vivian by reading her a simple children's story rather than reciting one of Donne's poems. It is as if Edson is deeply questioning the relevance of intellectual endeavor to "real" life, and saying that in the end what really matters is one's closeness to others and one's ability to give and receive compassion. Yet typical of this nuanced and thoughtful play is that Edson does not veer so easily into one predictable corner. In a flashback scene with her father, Vivian demonstrates the liberating force of language as she asks her father what the word "soporific" in her Beatrix Potter book means, and his answer "opens up a whole new world to her: the enchanted forest of words. . . . Your [the audience's] thinking about her becomes more complicated once you see how the very idea of knowledge makes her five-year-old face come alive."[37] Likewise, E.M. shows that the best use of learning is to help people. Her sophisticated understanding of literature is not there to be trotted out and displayed, to belittle others as Vivian does in her prime. Rather, it equips her for life, enabling her to adapt to situations, assess what people need, and more readily help them, as when she quickly sees that a children's story would give Vivian more comfort than anything else, even Donne. As she reads aloud *The Runaway Bunny* to Vivian, she seems to be the missing mother figure, complementing the father we had seen in the earlier flashback. Yet she also retains her professional persona, commenting to herself on the story's deeper meanings, bringing her eighty years of experience and accumulated literary expertise to bear on its simple utterances: " 'If you become a fish in a trout stream,' said his mother, 'I will become a fisherman and I will fish for you.' (*Thinking out loud*) Look at that. A little allegory of the soul. No matter where it hides, God will find it. See, Vivian?"[38]

Another of Donne's poems, Holy Sonnet Five, says essentially the same thing as this simple and classic children's tale: no matter where it hides, God will find the soul. Prefiguring the reading of *The Runaway Bunny*, this sonnet is given in its entirety during a recalled lecture by Vivian in the middle of the play. "But who am I, that dare dispute with thee? / O God, Oh! Of thine onely worthy blood, / And my teares, make a heavenly Lethean flood, / And drowne in it my sinnes blacke memorie" becomes central to the play's collective musings on death. The "Oh, God" utterance returns in the final scene with new meaning when spoken by Jason in an agony of remorse and shame for his physical violation of her (for the second time; the first was his internal pelvic examination of her, brusque and demoralizing, which is shown to the audience early in the play). In Vivian's interpretation of this poem as given in her recalled lecture, "the argument [of the poem] shifts from cleverness to melodrama, an unconvincing eruption of piety: 'O' 'God' 'Oh!' "[39] Emulating what she believes to be E.M.'s uncompromising approach, Vivian finds this melodramatic and weak. As she delivers this pronouncement, she is authoritative and intimidating, whacking the screen with her pointer to emphasize her words even while she is bald and dressed in her sacklike hospital gown—the old Vivian superimposed on the new. Her physical demeanor and her literary criticism bear the firm imprint of her tutor, the great E. M. Ashford, especially in the disdain for "histrionic outpouring," "hyperactive intellect," and "overwrought *dramatics*" in favor of simple acceptance of forgiveness. [40] As Vivian puts it, "True believers ask to be *remembered* by God. The speaker of this sonnet asks God to forget."[41] Theatricality is linked with insincerity and cowardice. No wonder Edson chooses to break the fourth wall and constantly remind the spectators that they are in a theater.

The emphasis on simplicity was already passed on to Vivian by E.M. in the flashback tutorial that has become, Vivian realizes now, a turning point in her life. She did not understand what E.M. was trying to tell her at the time; now Vivian sees that she tragically misunderstood her mentor, adopting E.M.'s toughness without her compassion, her rigor without her capacity for tenderness. Ironically, it is not this Donne poem but a children's story that has the final word about the immortality of the soul and the certainty of forgiveness—taking both E.M. and us by surprise.

Jason's unwitting and helpless echo of the "O God, Oh!" line is fraught with poetic meaning because of the skillful way in which Edson has set us up to make the connection between what he is saying and what we learned earlier from Vivian's lecture. This is similar to the painstaking way in which Frayn teaches us the uncertainty principle and complementarity in *Copenhagen*, layering anecdote upon anecdote to illustrate these concepts even while the actors orbit the stage performing the interactions of the particles they are describing. At this point in the play, the line should remind us of Donne and of the struggle for self-discipline: not to "hide from God's judgment, only to accept God's forgiveness"; or, in Jason's case, to realize that "overweening intellect" got in the way of basic humanity. Jason is the 1990s Ridgeon, a doctor who has stopped seeing the bodies he is treating as human beings and is interested only in their potential to further his own research. His repeated "Oh, God" articulates Vivian's point in her lecture on the sonnet: "The speaker of the sonnet has a brilliant mind, and he plays the part convincingly; but in the end he finds God's *forgiveness* hard to believe, so he crawls under a rock to hide."[42] As the stage directions indicate, Jason is figuratively crawling under a rock, curled up on the floor wishing to hide, for the duration of this scene. That he is able to say these words at all, though, seems to signal some hopeful change in him, too; perhaps he learns the same lesson as Vivian about opening one's eyes and hearts to the people around one, and ceasing to regard her as a mere vessel for his own success. He is able to show remorse just as Vivian has learned to do, and Edson seems to be saying that even if Vivian will never see that remorse, it is never too late, as Jason will perhaps be changed in his attitude toward his patients from now on.

The Neurologist's Tales

The popular writings of neurologist Oliver Sacks about the cases he has encountered have inspired not one but two acclaimed plays in recent years as well as some films. One play is Brian Friel's *Molly Sweeney* (1995), based on Sacks's story "To See and Not See"; the other is a "theatrical research" by Peter Brook and Marie-Hélène Estienne called *The Man Who* (2002), based on Sacks's story "The Man Who Mistook His Wife for a Hat." They will be considered here in terms of their theatrical emphases—the ways in which the playwrights use the scientific material for dramatic purposes.

Molly Sweeney is another "medical" play that features fraught doctor-patient relationships, an element added by Friel in his reworking of "To See and Not See." Sacks's piece recounts the true story of a blind man named Virgil on whom Sacks operated and restored sight. Virgil's life did not improve as predicted but actually deteriorated as he suffered setbacks associated with having to learn to see, to make adjustments that we take for granted. One of the fascinating truths to emerge from this account is that seeing is not second nature but must be learned just like other skills. Sometimes this process of learning is so painful and traumatic that the patient reverts to his or her previous methods of "seeing," for example, closing his or her eyes to identify a flower by its feel and smell rather than by looking at it. The patient enters a condition known as "blindsight." These are themes that Sacks has repeatedly returned to in his writing about neurology, from his renowned collection of case studies *The Man Who Mistook His Wife for a Hat* to a recent piece in the *New Yorker* entitled "The Case of Anna H."[43]

In *Molly Sweeney*, the basic story is kept intact, but the main character is a woman whose husband, Frank, a man of causes that he takes up enthusiastically and then drops when they fail to work out, persuades Molly—his latest cause—to have an operation to restore her sight. In their zeal to "correct" Molly's problem and gain success for themselves, the doctor, Mr. Rice, and Frank fail to predict—are blind to—the devastation they will cause. This once confident and happy woman loses her grip on her sanity when she acquires sight and is overwhelmed by the adjustments she must make. Once content and confident within her so-called disability, Molly becomes handicapped by the "gift" of sight.

One of the ways in which the play demonstrates this paradox visually is that the three characters never interact on stage—they literally do not see each other as they take turns telling their stories to the audience, each from his or her own perspective. They form a bond with the audience but not with each other. There is a tragic failure to see, and the play raises questions about who really has sight in this play, much as *Oedipus Rex* and *King Lear* do in having the physically blind characters show much greater insight than the sighted ones. In a diary that he kept while writing the play, Friel refers frequently to musical analogies in describing the emerging form of the play. "Maybe not distinct monologues; contrapuntal; overlapping," he thinks at one early stage of the writing process; the next week he notes "there are three voices—in fact the play is a trio for three voices."[44] Months later, despite radical changes to the evolving piece, Friel is still "casting the play in monologues; or duets; or trios; but not in any kind of usual dramatic exchange."[45] This musicality both haunts and eludes Friel, as he write despairingly weeks later on completion of the first act: "What is lost, so far, is the overreaching, perhaps excessive, notion that this could be a trio—all three voices speaking simultaneously, in immediate sequence, in counterpoint, in harmony, in discord. Instead I have a simple linear narrative in traditional form."[46]

Whether or not one agrees with Friel's severe self-judgment here about his failure to write contrapuntally, his aspiration to have the play mimic musical forms and his disparagement of "simple linear narrative in traditional form" seems utterly consistent with the kind of formal innovation that characterizes recent science plays. It also provides another example of the brilliant merging of form and content in science plays, for the emphasis on the sound of the play parallels a concomitant de-emphasis on its visual qualities—surely no coincidence in a drama about blindness. We do not have much to look at on the spare stage, populated by just three actors and three chairs (as in *Copenhagen*), and the lack of interaction among the characters gives us even less to watch. In short, "looking," "spectating," "observing," "watching"—all the synonyms for the act of seeing become less relevant than the sense of sound, as Friel successfully forces us to rely far more on hearing than on sight, just as if we were blind. The play attempts through its formal strategies to give us a taste of what it would be like to be Molly before her operation, and it is so subtly done that many audience members will not even notice the process taking place, only feel its effects as their playgoing experience eerily mimics the material they are "watching." The play thus calls on different faculties of observation from sight, and points up our overreliance on seeing as the primary way of sensing the world around us. Even a subtle change in the way Friel refers to the play in the diary over the course of a few months shows how he came to regard blindness rather than sight as its main theme: at the start of the play's gestation, Friel calls it "the sight play," then it becomes "the sight-blind play," then "the blind-sight play" and then finally and repeatedly "the blind play."[47] Blindness rapidly displaces sightedness as the play's central concern.

There is a long tradition of plays that treat the theme of blindness—from Sophocles' *Oedipus Rex* through Shakespeare's *King Lear* through to the modern period with Maeterlinck's *Les Aveugles*, Synge's *The Well of the Saints*, and Pirandello's *Henry IV*, for example. Despite his awareness of such literary precedents, in his program notes to the play "Friel himself refers us not to literature or to myth but to the scientific evidence of clinical psychology."[48] Friel takes the tradition in a new direction by incorporating the scientific underpinnings of the rich metaphor of blindness into the fabric of his play. Friel's diary about the process of writing *Molly Sweeney* reveals that while he was struggling with the musical elements of the dialogue, he was also having trouble integrating the neurological terms: "to balance the material in such a way that the medical element is essential but always subservient to the human. This means that the medical-technical processes and language have to be so thoroughly absorbed that all that is left is the watermark, the coloration."[49] Friel consulted numerous books on vision and blindness, but then found that they inhibited the creative process: "The blind play keeps getting snagged in complex medical explanations. . . . The play is about *people* and the medical condition of one of these people mustn't be allowed to dominate."[50] Even before these thoughts, at the very start of writing the play, Friel declares himself "weary of reading about prostheses, nystagmus, visual and tactile experience, etc. etc." and homes in on the play's emerging "fundamentals," namely: "a person is restored to sight[,] the experience is enormously difficult[,] the new world is a disappointment—the old world was better[, and] the person goes into a decline and dies."[51]

Friel wanted "a medical story that is also offered as a love/spiritual story."[52] He eventually found a balance between the medicine and the characters' lives. The play introduces "real" science by discussing terminology related to the neurology of blindness, such as "engram" and "agnosia," in depth as Frank recounts the research he has so enthusiastically done on Molly's condition. He explains these concepts directly to the audience through funny anecdotes that are also poignant in the way they underscore the teller's own inability to see the message hidden in the stories. One such example is Frank's anecdote about moving a badger sett, struggling to get it to a different location so as to save the badgers from flooding once the river reconstruction begins, but being foiled by the animals' stubborn refusal to accept this new environment. The parallel with Frank's forcing of the change of environment on Molly by having her undergo the operation is clear to the audience but not to him; even as he tells the story he remains oblivious to its metaphoric resonance, providing an especially poignant case of dramatic irony for the audience.

Friel makes several significant changes to the original story. First, he turns Virgil into Molly. Making the story about a woman adds an extremely interesting dimension that opens up several other possibilities to explore. This woman Molly is deeply dependent on men, so questions about gender in general surface. Her relationships to Mr. Rice and Frank are troubling, because she clearly regards them both as father figures, shown against the backdrop of her loving recollections of her deceased father, and they exploit her desire for such role models for their own purposes. Molly is all too eager to fill the gap left by her own beloved yet flawed father, who reared her while her mother was in and out of institutions for a nervous disorder but who never enrolled her in a school for the blind. Second, not only does Friel change the gender of the central character, he changes the personality quite drastically. Where Virgil is flaccid, passive, slow, and rather uninteresting, Molly is sparkling, vivacious, energetic—an altogether more colorful and engaging character. This has a tremendous impact on our connection to the respective characters, as it means that our identification with Molly and her situation is very strong and our attitude toward the doctor becomes more critical.

Mr. Rice, the doctor, adds to this by being harshly self-critical, calling himself the "rogue star" brought low by the "hunger to accomplish, the greed for achievement."[53] When Molly and Frank first came to see him, "I liked her at once," but he also had "a much less worthy thought": that he could "rescue . . . transform . . . restore" his career and reputation through a successful operation on her eyes.[54] The metaphor he most often invokes when talking about his own tragic past is Icarus and Daedalus, the overreaching father and son of Greek mythology, as if he is aware of his own hubris and the mistakes it brought about. His wife left him for his best friend, who then died in a plane crash. Rice fell into a "terrible darkness" that lasted for almost eight years.[55] He still drinks. He sees Molly as his ticket out of this awful slump. He can be "posh" and "icy"; Frank deems him "a right little bastard at times."[56] But we also find out that Molly warmed to him because "of all the doctors who examined me over the years he was the only one who never quizzed me about what it felt like to be blind—I supposed because he knew everything about it."[57] She trusts his ability to "see" her experience, in the fullest sense of the word.

The play questions our definition of and assumptions about disability. Molly truly shows what it means to be differently abled, not disabled. She refers to "her world" of blindness and shows us that it is a wonderful place not at all diminished in comparison with our own, even though, in one of the play's most painful moments, she calls it "that silly world I wanted to visit and devour" after the operation.[58] She is being conditioned to think of her blind world as "nonsense."[59] This causes the tragedy of the play: the withdrawal of this formerly well-adjusted, vibrant woman into her own fantasy world as she loses her job, her friends, her husband, and her sanity in her retreat from sight. Mr. Rice recalls, "It was hard to recognize the woman who had first come to my house. The confident way she shook my hand. Her calm and her independence. The way she held her head. How self-sufficient she had been then . . . so naturally, so easily experiencing her world."[60] She ends up following the fate of her mother, even winding up in her mother's old institution. "She had moved away from us all. She wasn't in her old blind world—she was exiled from that. And the sighted world, which she had never found hospitable, wasn't available to her anymore," Rice tells the audience. "My sense was that she was trying to compose another life that was neither sighted nor unsighted, somewhere she hoped was beyond disappointment; somewhere, she hoped, without expectation."[61]

The end for Molly is strikingly similar to the end for Vivian Bearing: each lies helpless and shrunken in her hospital bed, completely alone and physically and mentally reduced. Rice visits Molly and finds her sleeping, "propped against the pillows; her mouth open; her breathing shallow."[62] He is told that she "could last forever or she could slip away tonight." Molly's last monologue shows how completely she now lives in her fantasy world, inhabited by ghosts. She seems lucid, and then she tells us that she was awake when Mr. Rice visited her, and only feigned sleep while he took her hand and said, "I'm sorry, Molly Sweeney. I'm so sorry." Yet we do not know if this too is a fantasy, sandwiched as it is between "visits" from her dead mother and father. She continues to haunt Mr. Rice, who leaves his job and the community that he, Molly, and Frank had shared, just as Jason will surely be haunted by his failure with Vivian: his failure to treat her humanely while she lived and to let her go on her own terms, to "slip away," to her death.

In *Molly Sweeney*, just as in *Wit*, a female patient receives mistreatment from seemingly well-intentioned male doctors. As Friel tersely puts it, "The men force her to be sighted. The process kills her."[63] Brook and Estienne's *The Man Who* is very different in tone from either of these works, and does not seem concerned with gender in particular. In fact, it reads much more like the case histories on which it is based, and is less

concerned with individual characterization—with people and their stories—than with exploring scientific and medical findings.

Brook did not arrive at neurology as a subject and the hospital as a
setting right away. Looking to science as the new mythology for our times,
Brook immersed himself in "the fascinating world of quantum physics,
[which] seemed at first to be as rich and as disturbing as mythic India, but
soon I was forced to recognise that physicists as people are very rarely
as unusual as their discoveries, and charts, diagrams, calculations, and
computers are hardly the human material that theater demands."[64] Then
his wife gave him Sacks's book *The Man Who Mistook His Wife for a Hat*.
Shortly afterward Brook visited Sacks at his New York hospital and was
given a guided tour:

> Here I was amazed to find, in the field of neurology, a legitimate basis for
> theater work. I began to sense why the great Russian neurologist Alexander
> Romanovitch Luria had called neurology a "romantic science." Through-
> out a long and active life in the dangerously shifting conditions of Soviet
> Russia, Luria had maintained that a human science that was coldly factual
> was incomplete. Indeed, he was passionately involved in examining every
> verifiable detail of how the body and brain operate, yet this was not enough.
> The human element only appeared when each individual was accepted as
> unique. Seen in this way, science certainly becomes "romantic," and the
> inner landscapes of the brain do indeed suggest what in another mythol-
> ogy—the Persian poem The Conference of the Birds—is called the "Valley
> of Astonishment." But if neurology is romantic, it is also very exact. . . . A
> neurological patient has a specific lesion in a precise place, and as a result
> his behaviour changes. The neurologist has to develop to the finest degree
> his power of observation, as every tiny movement can be a clue to what is
> occurring somewhere, out of sight, in the brain. In the theater as well,
> behaviour is our raw material, so as the basis of our new work, we could at
> last return to everyday life, distilling ordinary, recognisable movements
> down to their essence and observing the irregularities that show a hidden
> tremor in the brain.[65]

As with his other work, Brook is concerned with the essence of what
makes us human. The difference is that in neurology he has found a new,
more exact and scientific source for excavating "the human element" in
all of us, as opposed to the vast and vague canvas of madness represented
by his now-famous piece *Marat/Sade*.

Conclusion

There is a tradition of plays about doctors and the medical profession that has been strengthened recently by the emergence of new plays treating the doctor-patient relationship in novel ways. From surgery to neurology to virology and endocrinology, and from syphilis to AIDS, medical plays cover a staggering range of illnesses and diseases. A unifying thread is the device of metatheater: breaking the fourth wall by having the actors directly address the audience, or using a "play within a play." Whatever form the metatheater takes, it draws attention to the fact of the audience being in a theater and prohibits the total suspension of disbelief. One of the reasons that *Wit* and *Molly Sweeney* succeed so brilliantly in conveying the science at their core and the nuanced and powerful doctor-patient interactions is their fluid movement in and out of realism by having the characters talk directly to the audience. This works extremely well in service of the text, far better, for example, than the more or less straight realism employed by earlier plays like *The Doctor's Dilemma* and *An Enemy of the People*, in which we sit and watch doctors and patients interacting as if we are invisible.[66] Yet why is this the case? Why should the simple, age-old device of breaking the fourth wall be especially effective for these recent medical plays?

It is not simply that the actors talk directly to the audience. It is a combination of this and a set of extremely deft theatrical metaphors in which form and content are seamlessly merged so that the one enhances the other. Vivian Bearing confides in the audience, laying bare her soul even while she invokes a poem about painful self-revelation whose theme is further enacted a little while later on stage by her doctor exposing his flawed judgment in the horrifying final scene. Molly, Rice, and Frank talk about sight and blindness to the audience, explaining all the nuances of each condition Molly experiences, yet never once interacting or "seeing" each other, perfectly demonstrating the tragic theme of the play—that we do not see or understand each other even when we think we do, while Molly in her blindness was the one who could truly see. In each case, the staging of the play—the play's theatrical strategies—helps to enact or perform its content in ways that the text alone cannot convey.

This is entirely consistent with the theatricality of contemporary science plays in general, from *Copenhagen* to *Arcadia* and beyond. In the doctor-patient plays, however, the combination of direct address and enactment of theme is even more powerful because the situations are so ripe for empathy, so moving and visceral: we are talked to by a cancer patient who is dying and who takes us through her experience in the hospital, and by a blind woman whose journey toward sight is harrowing because all her experiences, all the things she tells us about, are drawn from everyday life, and so they are things with which we can empathize. At a time of greater public awareness of terminal illnesses and concern on the part of doctors and patients alike as to how to deliver such news,[67] plays like these can directly contribute in valuable ways to our understanding of both perspectives. The cathartic element of these works is so strong because we are not just witnesses, watching the play as if eavesdropping on these lives, but are made to identify completely with these very human and universal predicaments: illness and death. Unquestionably the roots of such catharsis lie in the tradition of the operating theater and the anatomy lesson, as mentioned in chapter 1.

One might argue that doctor-patient plays are important to investigate not just as a key part of the tradition of science in the theater but because they tap into deeply held feelings: "Since all members of an audience must inevitably feel strongly about the profession that claims mastery over sickness and death, their interaction with theatrical presentations on medical subjects must reach deeply into their psyches."[68] A play that does this particularly effectively is Margaret Edson's *Wit*, which brings plays about doctors and patients to a new level by featuring a complex metaphor, a nuanced metatheatrical technique, and an unflinching honesty about the experiences of terminally ill patients and their doctors. The play's emotional and physical rawness, combined with its intelligence and wit, make it the prime example to date of successful plays about medicine. It also represents a successful bridging of the "two cultures" of the sciences and the humanities, since it exposes audiences to both areas in mutually enlightening ways: the poetry of John Donne set against the decidedly unpoetic atmosphere of a cancer ward. True to the nature of the best science plays, it shows how each culture can enrich the other.

8 "Just a Fiction":
Staging History and Truth

By now we have seen how science plays can provide an exciting way of unifying the "two cultures" outlined by C. P. Snow through the unique possibilities of the stage to bring disparate disciplines and viewpoints together before a live audience. Michael Frayn's *Copenhagen*, the release of the Bohr and Heisenberg letters, and the outpouring of commentary on the play from physicists and historians of science collectively demonstrate the effectiveness with which science plays can engage and challenge assumptions not only about historical events and people but also about their appropriateness as dramatic material. Other plays discussed in this book have likewise attracted attention, though on a more modest scale, as much for their depictions of real scientists as for their dramatic interrogation of the nature of scientific pursuit and the increasingly vexed ethical issues at the heart of scientific progress. In many cases, such attention has brought out significant and often heated controversy and debate between the theatrical and the scientific and historical communities. In fact, the issue of playwrights depicting real-life scientists, historical events, and scientific ideas on stage has become so central to the reception of science plays that it threatens to drive an even deeper wedge between the "two cultures" that have a stake on either side of the debate.

The urgency of this debate and the misunderstandings that have ensued warrant an in-depth analysis from the perspective not often represented in the debates between historians and scientists: the theatrical and literary viewpoint. This chapter will address the controversy over playwrights' use of history from the standpoint of theatrical performance, literary criticism, and theory, in an attempt to restore a sense of balance in the debates about science plays, to elucidate the main issues and conflicts surrounding them, and perhaps ultimately to remove the wedge that is threatening to divide the "two cultures" even further. The chapter will also address the problem of science plays' relation to postmodernism, since the two have been linked in criticisms of playwrights who allegedly misuse history on the assumption that in the postmodern world there is no absolute "truth." Here, again, *Copenhagen* will be something of a case study of science plays generally, as the debate it has generated continues to raise these concerns. The chapter will also discuss selected examples of recent plays by scientists and historians who have attempted to put science and the history of science on stage as correctives to what they see as the misuse of such material by playwrights.

The Responsibility of the Dramatist to Historical "Truth"

"How do you excavate the intimate moments of historical characters?" asks the playwright Lawrence in Timberlake Wertenbaker's play *After Darwin*.[1] As the author of the play-within-the-play about real historical characters (Charles Darwin and Robert FitzRoy, captain of the *Beagle*), Lawrence is asking a question that is directly relevant to actual playwrights like Kipphardt, Frayn, and Wertenbaker. One of the most persistent questions raised by *Copenhagen* is the degree of poetic license allowed the dramatist to use real characters and events. Is it all right to put words in their mouths, to speculate on what really happened, when the people and situations described have such historical impact and sensitivity? As Marko Juvan notes, since literature often functions as a repository of culture, this is an important question; "how, on the one hand, literature can be a foundation, even a means for constructing cultural (national, civilizational, racial, class, gender) identities, permanence and continuity, and on the other hand, a subversive force, a producer of incessant differences."[2]

Given this important role for literature within culture, many historians and scientists are deeply skeptical about the license they feel playwrights take with the truth. This is captured by the historian of science Lewis Wolpert in his entertaining playlet "Good Evening, Galileo," written to protest the way, in Wolpert's view, Brecht distorts history and science in his play. Wolpert interviews Galileo after they have both watched a production of Brecht's play, and finds Galileo displeased. "Do I really eat in so gross a manner? No, the real question is whether this is an accurate reflection as to what happened or has the author simply used my life to put forward a set of his own ideas? Once one uses real people in a play one must have some obligation, it seems to me, to depict them as accurately as possible. I find the title particularly annoying 'The Life of Galileo.' What a simplification; what a distortion; as if my whole life was nothing more than a battle with the Church."[3]

Brecht's point of view has been put forth in several ways, both in his copious writings on theater (a standard text, translated and edited by John Willett) and in the numerous works about Brecht. In the introduction to his translation of Brecht's *Galileo*, Eric Bentley addresses playwrights' use of history in great depth. He concedes: "If what playwrights are after is fiction, why do they purport to offer us history plays at all? A teasing question, not, perhaps, to be answered without a little beating around the bush."[4] Bentley draws on sources as varied as Aristotle, Pirandello, and Lukács to defend Brecht—and other "good playwrights"—using history and historical figures to fictional ends. "Drama has a different logic from that of fact. History can be (or appear to be) chaotic and meaningless; drama cannot. Truth may be stranger than fiction; but it is not as orderly."[5] He notes that "very much of our 'knowledge' of the past is based on fiction. . . . The question is whether the factual distortions [in a play like *Galileo*] have to be accepted at face value."[6] Bentley's main point is, as Lukáks would argue, that "while playwrights and novelists depart from the *facts* of history, they still present the larger *forces* of history."[7]

This is an essential point to remember about "history" plays like *Copenhagen*, *After Darwin*, *In the Matter of J. Robert Oppenheimer*, and *Galileo*. Introducing his play *In the Matter of . . .* , Heinar Kipphardt makes a similar argument, invoking Hegel to explain his balance of history and fiction. Kipphardt notes that his guiding principle is to expose, as Hegel says " 'the core and significance of a historical event by freeing it from the adventitious contingencies and irrelevant accessories of the event,' to 'strip away the circumstances and aspects that are of merely secondary importance, and to replace them with such that allow the essence of the matter to appear in all its clarity.' "[8] The advent of New Historicism has called such assumptions into question, however; it is precisely the matter that gets "stripped away" in the attempted distillation of a historical truth that most literary scholars and many historians would now find most valuable to investigate.

Most literary theorists would agree that a piece of fiction creates its own world, regardless of any reference it might contain to the outside, "real" world. "The literary utterance," writes Jonathan Culler, "brings into being characters and their actions . . . [and] literary works bring into being ideas, concepts, which they deploy."[9] Thus, literature involves "an active, world-making use of language" that "helps us to conceive of literature as an act or event. The notion of literature as performative contributes to a defense of literature: literature is not frivolous pseudo-statements but takes its place among the acts of language that transform the world, bringing into being the things that they name."[10] This is essential to remember when dealing with *Copenhagen* and other historically based plays. No matter how much its characters and events seem to derive from the real world, *Copenhagen*'s world is separate and fictional. "Objections on historical grounds," writes Bentley about Brecht's *Galileo*, "cannot be upheld, because the poet has created a vision that transcends literal reportage."[11] This could be said about any historically based play, including *Copenhagen*.

What happens when that fictional world clashes with the "real" world? As narratologist Marie-Laure Ryan points out:

> Though the fictional text creates an autonomous fictional world, this world can present some degree of overlap with the real world. . . . In historical novels [and, likewise, historically based plays like *Copenhagen* and *In the Matter of J. Robert Oppenheimer*], many of the propositions expressed by the fictional discourse happens [*sic*] to be true in reality; but this does not turn the text into a blend of fiction and nonfiction. *The primary reference world is the fictional world*, and . . . all the propositions expressed by the text yield truths about this world. But because propositions can be separately valued in different possible worlds, the real world may function as secondary referent. This explains the potential didactic and cognitive value of literature.[12]

Some of the hottest debates about *Copenhagen* have arisen out of this thorny issue of the real world taking a secondary role to the fictional one; many critics find themselves naturally inclined to view the real events and people as primary referent and the play's fictional re-creations of them as secondary.

This helps to explain the strategies at work in plays like *Copenhagen*. "The 'language game' of nonfiction is not defined by the objective relation of the text to the world, but by the rules that govern the use of the text, and bind sender and receiver in a communicative contract."[13] This contract is extremely complex in the case of *Copenhagen*, with many audience members having firsthand experience of the contexts surrounding the events Frayn deals with—some remember Bohr and/or Heisenberg personally, some worked at Los Alamos or are related to those who did, some, like Hans Bethe and John Wheeler, are actually mentioned in the play— and each one is personal and different from the next. "While the rules of the game specify that the text of nonfiction is to be evaluated in terms of truth, the text itself is unable to establish its own validity. The reader evaluates the truth value of the text by comparing its assertions to another source of knowledge relating to the same reference world . . . [for example] information provided by other texts. The need for external validation means that the nonfictional text stands in a polemical relation with other representations."[14]

One may quote this passage in relation to *Copenhagen* because its use of actual historical characters and real events has proved unsettling and caused a slippage between the terms "fiction" and "nonfiction" for many readers/viewers. Indeed, the physicist Gerald Holton has called it "fiction" in a pejorative sense: "Though it is a gripping drama, it is still a work of fiction."[15] The qualifying language of this sentence ("though . . . still") misses Frayn's reiterated and ironic point that his play was never meant to record history but rather to explore "the epistemology of intention." Such language also exposes a glaring bias against fiction as a whole (and by implication art itself), as if it is secondary to the writing of "factual" forms such as history and of science. This sort of hierarchical placement of one "culture" above another—in this case, hard science and historical "truth" above imaginative and creative art—seems outdated and beside the point.

Note the claims to authenticity in the introductions to docudramas like *Inherit the Wind* and *In the Matter of.* . . . These tend to seem overly emphatic and defensive in tone. In the case of the latter play, Kipphardt's prefatory remarks in "The Play in Relation to the Documentary Data" show him at terrific pains to point out that "the author adheres strictly to the facts which emerge from the documents and reports concerning this investigation."[16] His claim that "after mature consideration" he "deliberately confined himself to drawing only upon historical data for all the facts presented in this play" is somewhat undermined or at least qualified by the admission that "some filling-in and intensification was necessary in order to achieve a more tightly knit as well as more comprehensive documentation and, as such, more appropriate to the stage."[17]

In other words, the playwright wants to have it both ways: to preserve absolute authenticity and fidelity to the "data" and "facts," yet still manipulate them in the name of theatricality and drama. Thus Kipphardt's introduction to the play begins with this insistence on his adherence to documentary truth and ends with the startling revelation that he has in fact inserted completely fictitious monologues into the otherwise factual dialogue of the play.

One could argue that even if he did not invent any dialogue, even if the lines were verbatim as stated in the proceedings, the mere act of placing these factual texts in the mouths of actors and putting them on a stage alters the historical authenticity of the experience, because it introduces the element of mediation that performance involves. The audience knows it is watching a play, and that no matter how authentic that play claims to be, it is an interpretation, not a re-creation, of real people and events. In short, the debates that have raged around *Copenhagen* and other science plays have been led by scholars, critics, and experts in the relevant fields the play concerns itself with. "Regular" audience members seem to have had no trouble separating fact from fiction and keeping Frayn's stage world distinct in their minds from the world of real events and people.

"Felicity," "Efficacy," the Theater/Performance Dichotomy, and the Science Play

In previous chapters, we have seen how J. L. Austin's ideas about speech-act theory and the concept of performativity to which it directly relates have been important to understanding how science plays work. The dismissal of *Copenhagen* (or any other historically based science play) as just a "fiction" is strangely similar to Austin's attitude toward literature. As J. Hillis Miller points out in his invaluable analysis of speech-act theory and literature, Austin distrusts literary speech because it tends to be "infelicitous." That is, one can never be entirely sure of the nature of the speech in a poem, play, or novel. This is largely because of its dependence on an audience to receive it, and how literary language is taken makes a tremendous difference to its meaning. In addition, literary speech does not "do" much; it does not effect anything in the same sense as speech acts in the real world. Hence Austin's distrust of it. "The firm exclusions of literature echo throughout *How to Do Things with Words* with stern but slightly desperate frequency," writes Miller, noting that despite the effort to exclude it, literature haunts the book, "vitiating the attempt to establish the conditions of a felicitous performative and so constitutes another kind of failure."[18]

In both cases—Holton on *Copenhagen*, Austin on literature in general—the critic displays an uneasiness about the nature of the object under consideration. It is as if, placed within a literary context, language suddenly becomes more slippery, less "felicitous" because less efficacious, less able to accomplish concrete tasks à la true performatives. How interesting, then, that *Copenhagen* has been accused both of talking too much while not showing enough (not being good theater, in other words) *and* of being "just" a fiction.

The fact that it is a play is, again, more than coincidental with regard to such assessments. As Stephen J. Bottoms argues, the rise of performance studies, with its emphasis on efficacy and immediate social change, has caused theater to be seen by contrast as quaint and old-fashioned, and purely for entertainment. "The notion of 'performativity' is celebrated as efficacious—it is a *potent* word, a *doing* word . . . —while 'theatricality' continues to be associated, unthinkingly, with ingrained connotations of empty show and ostentation, lacking in transformative potential."[19] Critics who dismiss *Copenhagen* as fictional are thus playing out this double condemnation of theater as both socially ineffectual (indeed, decadent) and a pale imitation of real life.

Perhaps more productive is the historian of science Robert Marc Friedman's approach to the issue of playwrights' use of history, both in relation to *Copenhagen* and in a more general sense:

> We must accept a basic truth: Theatre cannot depict comprehensive narrative history and ought not to try. Still, theatrical drama can stimulate thought and raise questions. No medium can better convey the immediacy of emotions—and science entails not only cold logic but also hot passion. . . . The playwright's skill, imagination, and intellectual grasp allow opening windows onto aspects of science history that can prod members of the audience to reflect or read further. The payoff might well be measured in the degree by which particular scientific events along with the persons linked with them can be transformed into public and dramatic property. Significant chapters in science history have a right to enter our cultural heritage and not remain merely the property of historians and scientists. A play such as Michael Frayn's *Copenhagen*, which although primarily concerned with the epistemological problem of what we actually can know of persons' motives, has clearly brought significant chapters in the history of modern physics to the attention of many non-specialists.[20]

In a similar vein, several prominent playwrights recently shared their thoughts in a *New York Times* article by Jonathan Mandell on the subject of dramatizing history. Mandell begins with a key point about the educational preparedness of audiences for such plays: "As social critics often lament, this is an age when many people get their knowledge of the past (and often the present) almost exclusively from their entertainment."[21] Does a play like *Copenhagen* thus have an even greater responsibility to educate people correctly, if so few audience members actually know enough about physics and history to exercise their own judgment and to understand the fine line between "fact" and "fiction"? This is indeed a powerful argument. If audience members already knew their quantum physics and had a stronger grasp of who Bohr and Heisenberg were, and what happened during World War II as nuclear weapons were being developed and deployed, would there be less of a burden on the playwright to render the events and people he or she is depicting as accurately as possible? This question troubles many critics of the play, who view the whole enterprise of dramatizing science and history with skepticism precisely because of the danger of distorting facts. If the play is indeed educating audiences, filling a hole in their knowledge, then is it not the duty of the playwright to get it right? "If real history and real people are portrayed, how accurate is accurate enough in plays? . . . Are there good and bad reasons to change the facts? When reaching back into history, do artists have a responsibility to more than their artistic vision?"[22]

The answer would be yes, if Frayn had started with the idea of educating his audiences about science and history. But his use of science is for thematic purposes. He uses the uncertainty principle and complementarity to make an analogy with human behavior, as a richly metaphoric and evocative way of probing the problem of intentionality. This is painstakingly elucidated not only in the play itself but in both postscripts Frayn has written. Critics of the play sometimes question the applicability of these principles of physics to human beings in general, finding it too easy and reductive. Yet according to Stephen Greenblatt, that is precisely how art makes its impact. "In any culture there is a general symbolic economy made up of the myriad signs that excite human desire, fear, and aggression. Through their ability to construct resonant stories, their command of effective imagery, and above all their sensitivity to the greatest collective creation of any culture—language—literary artists are skilled at manipulating this economy. They take symbolic materials from one zone of the culture and move them into another, augmenting their emotional force, altering their significance, linking them with other materials taken from a different zone, changing their place in a larger social design."[23] The manipulation of symbolic materials is not only the prerogative of the artist; it actually serves a vital cultural purpose in enabling greater understanding of our world. Greenblatt notes that we must ask

> fresh questions about the possible social functions of works of art. Indeed even if one begins to achieve a sophisticated historical sense of the cultural materials out of which a literary text is constructed, it remains essential to study the ways in which these materials are formally put together and articulated in order to understand the cultural work that the text accom-

plishes. For great works of art are not neutral relay stations in the circulation of cultural materials. Something happens to objects, beliefs, and practices when they are represented, reimagined, and performed in literary
texts, something often unpredictable and disturbing. That "something" is
the sign both of the power of art and of the embeddedness of culture in the
contingencies of history.[24]

This is another way of saying that the semiotics of theatrical performance
are especially complex: on stage, "something happens" to objects whose
meaning in everyday life might seem straightforward and clear. The same
could be said of stage language. Meanings become destabilized and inter-
rogated when placed in the framework of performance. This is precisely
what divides the theatrical viewpoint from the historical and scientific
with respect to the depiction of real people and events; in the former view
the destabilization is liberating and is the whole point, while in the latter
it is deemed to be inaccurate and thus threatening to the idea of truth.

Arthur Miller tells Mandell about writing *The Crucible* with the aim
of exploring not just actual, public events but the emotional states and
experiences of the characters in their historical situations: "What their
internal psychology was, was not on any record. . . . That is what I had
to create."[25] Frayn says essentially the same thing both in his postscripts
to the play and in any public discussions of it: he talks simply of trying to
get at the thoughts and intentions and motives of the people involved.
Both he and Miller are creating an "imaginary reconstruction," in Miller's
term. "It wouldn't be a document you would turn to for absolute historical
truth, if there is such a thing."[26]

It is precisely this clause that gets at the heart of the debate about
Frayn's use of history. It serves as a response to critics who have complained, for example, that Frayn's Margrethe bears no relation to the real
Margrethe Bohr, who was a model of vocal restraint and would never say
the sarcastic and biting things that pass the character's lips in the play.
Yet, to complicate matters even further, within literature itself there is an
ongoing debate about previously accepted distinctions between the genres
of fiction and nonfiction. "To the postmodern mind—as to any self-re-
specting avant-garde—the most entrenched of cultural categories offer
the most enticing targets," writes Ryan. "A case in point is the distinction
between fiction and nonfiction. . . . Postmodernism has placed nonfiction
in the hot seat."[27] Those who want to maintain a distinction between
fiction and nonfiction "will have to establish the legitimacy of nonfiction
in the face of anti-realist and relativist arguments."[28] That *Copenhagen*

has generated so much debate about its use of history surely relates directly to this fundamental problem separating the humanities and the sciences: how each "culture" approaches the issue of truth and reality and distinguishes "fact" from "fiction."

"To what extent can theater be used as a vehicle for exploring the history of science?" asks Friedman in the opening of an article for *Physics World*—for a readership of scientists—titled "Dangers of Dramatizing Science." Friedman's article is based on his presentation at the symposium on *Copenhagen* at the Niels Bohr Archives in Copenhagen in 2001; he also published these concerns in a similar article in *Interdisciplinary Science Reviews* in 2002. While Friedman is enthusiastic about Frayn's play generally, and an aspiring author of science plays himself, his keenness is tempered by these concerns about how playwrights appropriate historical characters and events, particularly if they are not well known by the audience.

Yet to use theater "as a vehicle for exploring the history of science" is one thing; to use history to create theater is another. Let us take a closer look at what Friedman has said about this issue, since he is both a historian of science and a playwright interested in bringing the history of science into the theater. Friedman seems to assume that Frayn is interested in "exploring the history of science," with theater just the vehicle for doing that, whereas really it is the other way around: Frayn uses the history of science as a vehicle for the theater. He is exploring what is essentially a new theatrical experience by opening up questions derived from his scientific metaphors (uncertainty, complementarity) about the epistemology of intention and the imprecision of human memory. What has been so frequently misunderstood about *Copenhagen* is that it does not in fact use history as an end in itself, or theater as the springboard for examining historical questions, but instead it takes history simply as material for creating theater that does what art in general does: poses questions. "Audiences may leave the theater with a wide range of impressions," cautions Friedman.[29] Among them is, of course, the impression of Heisenberg as sympathetic, and as Bohr's moral equal. Yet is this not precisely the point of art, to allow "a wide range of impressions" and stimulate the audience to think broadly about what it has seen? Friedman worries about the play getting away from "the playwright's intention,"[30] yet such concern seems beside the point after decades of literary theory have successfully demonstrated that intentionality of that nature is a naive concept. We have moved far beyond that, yet critics of *Copenhagen* are still adopting this argument.

Further on Friedman maintains that whereas historical drama generally deals with "characters who have already achieved legendary status: kings, queens, statesmen and others who are already public property," drama based on the history of science suffers because "few scientists . . . have achieved such status. Most scientists are not only not legendary but are hardly known to the public."[31] This makes little sense and seems to betray an old-fashioned attitude on the part of the author; these days one can possibly think of more famous scientists than statesmen, a result of the rise to prominence of science, and of our living in the "scientific age." Surely Einstein, who features in numerous successful plays, and Galileo, likewise a theatrical mainstay, are nothing if not "legendary."

Friedman concludes by rephrasing his main reservation: "When a playwright breathes life into a name from history and creates a seemingly real person who is as new for the audience as any fictional character, there should be some sense of responsibility for how that person is portrayed."[32] Presumably Friedman is referring to Frayn's portrayal of Heisenberg, since that is the most contentious aspect of the play. This sounds a great deal like the argument of Paul Lawrence Rose, one of Frayn's most outspoken critics, who maintains that Heisenberg was unapologetically supportive of the Nazis and worked willingly for them to try to develop atomic weapons before the Allies. Thus, to depict Heisenberg in a sympathetic light is, in Rose's view, to be anti-Semitic. Yet one could also maintain that to suggest as Friedman does that Frayn compromises the "intellectual and moral integrity" of his play by his depiction of Heisenberg is to condemn art for posing multifaceted ways of seeing characters and events, whether they are "real" or not. Surely that is the great value of art, and why we need it now more than ever.

It is worth dwelling on what Friedman has had to say as a critic because his views represent some of the most hotly debated aspects of science and theater, and because he is also an aspiring science playwright and therefore offers a rare case of complementarity. He has written a play called *Remembering Miss Meitner* about Lise Meitner, the codiscoverer (with Otto Hahn) of fission. Hahn received a Nobel Prize for his discovery, but (as in the case of Rosalind Franklin and the discovery of DNA) Meitner's important role was overlooked. Friedman's play is modeled almost exactly on the theatrical vision Frayn has devised in *Copenhagen*. It is set in "the chamber of historical record" long after Meitner and Hahn (and the third character, Swedish physicist Manne Siegbahn) are dead—precisely the kind of afterlife Frayn imagines for his play. Both plays deal with physics and nuclear weapons and refer to the same time period; although in *Copenhagen* Margrethe is the wife of a physicist, not one herself, she is intimately involved in Bohr's professional life, since she types his numerous drafts and serves as a sounding board and adviser. Both plays are peopled by three characters only, one female and two male. Both are heavily textual as the characters discuss their actions, words, and motives. And each play is concerned to answer a central question: In Frayn's case, why did Heisenberg come to *Copenhagen?* In Friedman's, why did Otto Hahn not give Meitner credit for her role in helping to discover nuclear fission?

In all these respects the two plays are similar. The great difference is in how each author handles his central question theatrically. It is striking that neither play provides a definitive answer to its main question. Frayn presents three successive "drafts" or possible scenarios and leaves the audience to choose which it prefers. Although no easy answer has been provided, the audience leaves having been through a process of discovery and experiment. Friedman's play never moves forward, returning exactly to where it began. There is no forward movement, no sense of growth in understanding for either the characters or the audience. The beauty of *Copenhagen* lies partly in the way Frayn paradoxically achieves some kind of resolution despite nothing being resolved, because of the extraordinary suggestiveness of the blend of science, history, and theater. By contrast, Friedman's play helps to illustrate the problem of many attempts at writing science plays. In a review of Friedman's play (or a preview version of it), Anders Barany summarizes some of Friedman's own arguments about putting the history of science on the stage. Barany writes: "On the one hand, the factual story can certainly be told in a direct way, leaving out answers to questions about 'riddles' or 'betrayals.' Such plays can work as an educational tool, but probably only for audiences who really want or need to be educated. On the other hand, the interpretational story can be told so far from reality as to make the play truly fictional. This might make it work as entertainment for normal audiences, but could frustrate those who happen to know something of the real characters portrayed. To steer clear of both extremes is not an easy task."[33]

Barany's use of the term "direct way" is interesting because it implies that some mediation on the part of the artist is what makes art; put across "directly," without such mediation, the play will have greater adherence to historical fact but probably less aesthetic merit. Translators confront a very similar situation in their work. They must make a choice: to translate with word-for-word literalness, with the boon of total fidelity to the original but the drawback of often plodding or "unartistic" language in translation; or to try to capture the essence of the original so that even if it is not a literal translation the reader/audience member will get the same meaning as the author conveys in the original, yet mediated through the translator's expert adaptation. Frayn, an accomplished linguist and an expert translator of Chekhov, knows well the nuances of translation and adaptation, and *Copenhagen* shows how that understanding can be applied to the translation of "real-life" material into a fictional world.

We have considered how a historian of science has tackled the problem of staging "real" events and people from the history of science; let us now look at how a scientist handles the challenge of staging scientific subjects and ideas. Carl Djerassi is a chemistry professor and the inventor of the birth control pill. In recent years he has turned his attention to writing plays and in fact states that he now considers himself to be a playwright.[34] He has written several plays with scientific themes, including *An Immaculate Misconception*, *Oxygen* (with Roald Hoffman, a Nobel Prize–winning chemist), and *Calculus*. All these have been produced in professional theaters and translated into numerous languages. This group of plays are all part of Djerassi's self-described "science-in-theater," which he has explained as having substantial scientific content and an unabashedly didactic aim: "What is wrong with learning something while being entertained? Or from the playwright's perspective, why not use drama to smuggle important information generally not available on the stage into the minds of a general public? . . . For my purpose, it is not sufficient to simply insert here or there some science into a play or have some characters who are more than just Frankensteins, Strangeloves, or idiots savants."[35] Djerassi has "smuggled" science into several of his plays. For example, in *Oxygen*, he and Roald Hoffmann dramatize the events leading to the discovery of oxygen in the eighteenth century and focus on the contributions of Antoine Lavoisier, Joseph Priestley, and Carl Wilhelm Scheele. These historical scenes are interwoven with scenes depicting contemporary scientists, namely, the Chemistry Committee of the Royal Swedish Academy of Sciences, who are trying to decide on the posthumous recipients of the first "Retro-Nobel" award. The play was hailed by scientists: Susan Greenfield called it a "brilliant example of a new genre: 'science-in-theater,'" Oliver Sacks wrote of it as "an extraordinary tour-de-force," and Murray Gell-Mann praised the way *Oxygen* "employs ingenious dramatic devices to explore the multiple facets of scientific discovery."[36] On the other hand, theater scholars have criticized Djerassi's plays for heavy-handed didacticism and for lack of character development, particularly when it comes to women, who in Djerassi's plays tend to come across negatively—the modern women ruthless and self-absorbed, the historical ones figuring stereotypically as mere supports to their husband-scientists.[37] In her review of the play for the *Irish Times*, Suzanne Lynch also points out that the production of *Oxygen* crystallizes "the one perennial difference between science and the humanities: money." Among the production's sponsors when it came to Riverside Studios in London were the "industrial giants" Dow Chemical Company and Pfizer, as well as a U.S. investment banker "who has swiftly acceded to the role of co-producer."[38] Lynch writes: "It is understandable, then, that some may resent what they see as a whimsical social experiment by scientists with a crusade to educate the scientifically illiterate masses, in an industry in which hundreds of plays by talented writers are rejected every year because of financial exigencies. This is the problem. It is the science rather than the drama of *Oxygen* that has attracted funding."[39] However, Lynch makes an astute final assessment of the play *Oxygen* that serves well to summarize Djerassi's science-in-theater contributions overall: she calls the play "important, nevertheless, even if its significance resides more in its generic achievement than in its qualities as drama. It marks a bold attempt on the part of science to take its place in the world of the humanities—and to exploit the potential of theater to bring scientific knowledge to a wider audience."[40]

Whatever the final verdict on Djerassi's plays, it is clear that he has made a significant contribution to the debate about the use of history and science on stage. He has also sparked a major controversy between members of the theater community on the one hand and the scientific community on the other. In terms of his aim of using theater to educate the general public about science, Djerassi's plays have enjoyed success, particularly those recent works he has written since turning his attention to "the school classroom," which in his own words is "the place where new pedagogy is best suited."[41] In terms of his aims in writing for the commercial stage, it is ironic that having started out trying to bridge the gap between the two cultures, this scientist-playwright has succeeded in widening it.

The last word on the controversy over plays using real historical events and people and depicting actual science should perhaps go to Kenneth A. Ribet, a professor of mathematics at the University of California at

Berkeley, who reviewed the recent math play *Partition*. This play takes
history and science and mixes them up gleefully, yet maintains accuracy
in its use of mathematics and its biographical references to real mathematicians and their work. Ribet noted that mathematicians were "disturbed"
and "startled" by the play's "historical distortion" and deliberate anachro-
nism: "In order to enjoy the play, one must relax the implicit identification between the historical Hardy-Ramanujan and the characters on
stage. Theater-goers who have little problem observing a goddess in discussion with a seventeenth-century mathematician on stage can make
their peace with a historical distortion that allows the audience to hook
up with a familiar and famous problem [Fermat's last theorem]. Once I was
able to separate the real Hardy and Ramanujan from their counterparts on
stage, I found only good things to say about *Partition*."[42] One can hardly
find a better way of punctuating this discussion of the use of history and
science on stage.

Reenacting and Shaping Cultural Memories

In addition to educating us in specific areas, literature also plays a central
role in our conceptualization of cultural memories. Jonathan Culler notes
the usefulness of "the model of the performative" because it "offers a more
sophisticated account of issues that are often crudely stated as a blurring
of the boundaries between fact and fiction. And the problem of literary
event, of literature as act, can offer a model for thinking about cultural
events generally."[43] This is partly why *Copenhagen*'s use of history is such
a sensitive issue: it causes us to reconsider the inherited cultural memories
of the race to develop and use atomic weapons, the consensus about who
the "good guys" were and who the "bad guys" were, and where Bohr and
Heisenberg fall on such a simplistic spectrum of good and evil. Whole
cultural and national identities are involved here, not just personal and
subjective accounts of history.

Plays like *Copenhagen* not only provide new insights into the production and dissemination of cultural memories but also represent a kind of
"third culture," a tentative refuge in the continuing clash between the
two cultures. *Copenhagen* is not just about cultural memory in the sense
of nationhood or ethnicity or the Allies versus the Nazis; it is also very
much about how we conceptualize the relationship of science to art. Frayn
brings the two together not by "dumbing down" the science or making
the form tangential to the content but by using the resources of the stage

to illustrate the complex science itself. Frayn's approach to truth uses science both accurately and metaphorically, and this in itself is unsettling because it juxtaposes two seemingly incompatible worldviews.

As I have noted elsewhere in this study, physicists have been borrowing from the vocabulary of theater to convey their ideas. In addition to the scientists, historians too are borrowing from the theater. "In comparing the work of the docudramatist and the academic historian," writes Attilio Favorini, "we may now observe that in endeavoring to represent the truth the documentarian simulates the historian's methodology and protocol: collecting data, citing primary source materials, producing charts and illustrations, employing real-seeming film clips and, via (e.g.) the Voice of the Living Newspaper or the invisible camera, approximating the desubjectivised, omniscient, third-person voice of the historian."[44]

This description of the historian's and documentarian's methodology might equally well characterize that of the scientist, on which it is of course founded. Scientific writing cultivates and demands a "desubjectivised" and omniscient voice. Much of its authority lies precisely in its mastery of such a style. But this is not the main point; what is significant is the deeply symbiotic relationship between historian and theater maker. As Favorini notes, "Inversely, and problematically, an important body of contemporary historiography describes the heavy reliance of the academic historian on the rhetorical strategies of the fiction-maker and theater artist." Favorini refers to Herbert Lindenberger's comments on the historian's affinity for "the theatrical metaphor to affect [sic] a sense of continuity (plotting) and dignity, to create with theatrical flair the 'panorama' of history."[45] How fascinating that he uses the term "dignity" to describe the desired effect of theater, to describe a quality of the theater that historians want to borrow. It is ironic but true, says Favorini, that "when the historian wants to create the impression of truth s/he draws on the techniques of theater; and when the documentary playwright has the same objective, s/he copies the ceremonial forms of the historian."[46] Many historians and scientists have quite the opposite view of the stage, particularly as a venue for scientific ideas in plays, feeling that theatricality detracts from or compromises objective truth and historical fact. The very notion of fiction that theater naturally implies seems problematic for some.

In a speech to a primarily scientific audience, Richard Holmes, biographer of Coleridge, addressed the problem of bridging the gap between the two cultures. He suggested that since the nineteenth century "science has 'resisted the metaphor' . . . Scientists too often stopped dead at the end of their inquiries and discoveries because they regarded the suggestiveness of these as none of their business, as 'not science.' " The bridge between the two worlds of science and art lies, Holmes argues, in this "imaginative widening" of the science into metaphors of great common applicability.[47] This is precisely what so many science plays, especially the contemporary ones, attempt to do. Whether they end up bridging or widening the gap remains to be seen.

Conclusion
••••••••••••••

Alternating Currents: New Trends in Science and Theater

At this point one might conclude that the science plays that have proved most memorable and significant have been those like *Copenhagen*, *Galileo*, and *After Darwin* that depict real events and people, almost always provoking controversy about the use of history in a fictional mode; and plays like *Arcadia* that are entirely fictional and engage complicated scientific subjects in brilliantly inventive and appealing ways. These are some of the works that make up the recent wave of science plays over the last two decades, and that collectively constitute a phenomenon of science playwriting, one that invites some attempt at taxonomy. We have seen how many of these recent science plays have sparked controversies and debates between prominent historians and scientists, on the one hand, and theater practitioners and scholars, on the other, debates whose high profile has tended to dominate the reception of science in the theater.

However, as this chapter will argue, most of these well-known, high-profile plays—in particular *Copenhagen* and *Arcadia*—are groundbreaking in their use of science but still rather mainstream in their theatricality. A second wave of recent science plays is quietly changing the way we think about science on stage, linking performance techniques and science in innovative ways that move away from the literary and historical foci of works like *Copenhagen* or *After Darwin* and reclaim the theatricality of the science play. It is this host of "alternative" science plays that I want to examine here. Simon McBurney and the Theatre de Complicite's *Mnemonic*, John Barrow and Luca Ronconi's *Infinities*, and Jean-François Peyret and Alain Prochiantz's various collaborative productions, particularly their Darwin trilogy, represent such "alternative" science plays. My intention is not to set up a simple opposition between such works and the more mainstream variety of science play but to explore their differences and especially how they utilize science in relation to the audience. These new pieces are being devised by contemporary theater practitioners who find that science plays often place too much emphasis on traditional modes of drama and dramaturgical strategies, and who present alternative approaches that try more directly to engage the audience with the ideas themselves. In what follows, I will look at some specific examples of these new kinds of science plays, what they do that is different from more mainstream science plays, and what they signify for future development of the interrelationship of science and theater.

Textuality

The main point of divergence that I want to stress is the notion of "mediation," which is the Italian director Luca Ronconi's term. As Ronconi argues, most "science plays" focus too much on biography, rendering the scientific ideas through the life of the scientific character. A play like *Copenhagen* does engage science deeply and meaningfully, but the science is mediated through biography. This gets the audience too wrapped up in plot and personalities: "A scientific topic such as the uncertainty principle has become part of Heisenberg's biography, a discovery such as that of radium has changed into the story of Madame Curie," says Ronconi. "So Madame Curie and Heisenberg have become the protagonists of plays, on the same level, so to speak, as Hamlet, Othello or Romeo and Juliet. Thus they have become real dramatic characters and the language used to portray them has no longer anything to do with their discoveries, it is the traditional language of drama."[1] Ronconi and others propose instead to build on the interest in performance and in new ways of presenting theater:

> The dramaturgy of the last few decades has shown us that there are other ways of writing for the theater by exploiting, for instance, the expansion (or the contraction) of space and time, which have both become elements of drama exactly like such traditional features as dialogue and character. This implies that a certain topic, let us say drawn from science, but it could be from high finance or economics, can enter the theater not disguised as usual, but as it is, in all its asperity and difficulty. From this new viewpoint we decided to try and see if we could find a meeting point, halfway between drama and science, neither totally on the side of drama nor totally on the side of science, neither something completely "formal" nor a kind of "popularization."[2]

In Ronconi's view, getting away from the traditional emphasis on character, and on other dramaturgical staples like linearity and causality, proves hugely rewarding because it allows the ideas to breathe and to dominate. This in turn allows the audience to see the impact of certain discoveries not in a personal, private sense but in the context of the encounter between ideas and people, between science and culture.

Both Ronconi and the French director Jean-François Peyret have used math and science in the theater in innovative ways, eschewing plot and character in the traditional sense of those terms and seeking instead to immerse the audience in the scientific ideas. In their engagement with science, they seek to create plays that seem less "mediated" than plays that explain the science through character and plot. Peyret wants his audiences to dream, a notion that is central to his work with science: "We aim to invite you this evening to give yourself over to dreaming, to enter the process of dreaming" in and about evolution.[3] This idea of the dream effectively captures the way in which the audience should experience the science. There is less dependence on the textual conveyance of ideas, more emphasis on the visual and the physical experience of the audience. Rather than employing literary devices or even such standard methods as plot and characters, the science goes directly to the audience, immersing them in the ideas theatrically. It is experiential rather than intellectual. In other words, the experience of the audience becomes much more about imbibing, sensing, or "dreaming" the science through its enactment than about listening to explanations of it from characters and following their stories.

As we have seen, it is certainly true that some of the most successful science plays employ a particular scientific idea or concept as an extended theatrical metaphor—they literally *enact* the idea that they engage. In *Copenhagen*, Heisenberg's uncertainty principle is the thematic center of the play, demonstrated through a series of speech acts that illustrate the position and momentum of observer and object through deft analogy with the actors and their characters, who all the while are moving about the stage in such a way as to enact the ideas of uncertainty and complementarity. Through dialogue composed largely of speech acts the science is both performed for us and transformed into metaphor on the stage, in a way that only the liveness and immediacy of theater can bring about. The dialogue does not merely reflect the principle; it makes it happen, with the audience participating in that act of creation.[4]

Now, however exciting this merging of form and content is, one could argue that *Copenhagen* is still conceptually quite traditional. Heavily textual, the play uses science, biography, and history to frame its main concerns, such as the problem of intentionality and the inability both to act and to observe oneself simultaneously. Jean-François Peyret calls *Copenhagen* "faux théâtre," or fake/false theater; it is "théâtre de la morgue."[5] He feels that plays like this are for the museum. Ronconi expresses similar feelings about *Copenhagen* and about Peter Brook's neurologically based production *The Man Who*, a "theatrical research" written with Marie-Hélène Estienne and based on the neurology case studies of Oliver Sacks. For all the input of the director and designers, there is no question that the text drives these productions. *Copenhagen*'s text is what has stimulated so much discussion among scholars in the fields of science and history, while its theatricality has received little attention. But in Peyret's view, "If they want to know whether Heisenberg was good or bad, they have access to the scientific debates if they want to, they don't have to come see a play. We don't have to do night school."[6]

Theatre de Complicite's founder, Simon McBurney, acknowledges that "everything begins with a text of some sort,"[7] and this is true of his work and that of Ronconi and Peyret. The difference is in the role and utilization of the text. In "alternative" science plays, there is no monolithic, cohesive, and unchanging central text. Often the theatrical production derives from a collection of ideas, a set of writings that serve as a springboard for further workshopping, rather than a fixed and final, stable, and authoritative script that is simply to be acted. So, while *Mnemonic* does have textual starting points such as the work of neuroscientists (quoted in the program for the play) and the dialogue spoken by the actors, there is much greater dependence on "communal storytelling that relied on the expressive powers of the body and the transforming capacity of inanimate objects."[8] Complicite's theatricality, with its very physical emphasis and the imaginative transformation of props, is very close to the acrobatic physicality of Peyret and Ronconi's work. The play moves from science on the microscopic level—the chemical processes in the brain that create, store, and activate memories—to the macro-level of how time works in the universe, and some of the most powerful moments in the production are not generated by a text. *Mnemonic* directly engages the audience—indeed, performs an experiment with the audience as the subject—through its interactive opening exercise of memory recall that sets up the framework for the rest of the performance. The production has drawn some criticism for the "universalizing" implications of its meditations on collective and individual memory. " In spite of attention to cultural specificity and difference, a human equivalency emerged as the dominant premise of the evening. . . . The play is from and about the new Europe. It was co-produced with the Salzburg Festival, and is designed for an international, or at the very least, a European audience. A fascinating opportunity has opened up in the arts for refashioning Europe through artistic and cultural practices, but this same opportunity is going to have some landmines as well," writes Janelle Reinelt in a review of *Mnemonic*.[9]

Taking the Plunge into Science

Just as *Mnemonic* immerses the audience in certain aspects of neuroscience and biochemistry, *Infinities* successfully engages mathematics by staging various "thought experiments" on infinity in theatrically diverse and innovative ways and in a highly unusual, interactive (some would argue site-specific) space. The audience is led through five different scenarios on the idea of infinity—from Hilbert's Hotel Infinity to Borges's infinite Library of Babel to a place where no one dies to a society capable of time travel—all performed in various spaces within the cavernous old warehouse in Milan's Bovisa area that once housed sets and costumes for La Scala.[10]

Ronconi makes few concessions to audience expectations of realism. He prefers to rely on more abstract elements such as symbolic movement: straight lines (actors moving perpetually forward and backward) or circles (actors sitting or walking endlessly around) that perform infinity. He also avoids realism by shielding many of the actors in gray wigs and white half masks (leaving the mouth and eyes visible but grotesquely exaggerating the cheeks and nose). *Infinities* places exhaustive demands on its actors. They often have to rush from one scenario to another, and whether hanging upside down from the ceiling or leaping across a vast expanse of books, they need to have the agility and stamina of acrobats. This is highly visual and physical theater. Actors also take turns rotating roles throughout the run of the show to avoid becoming complacent and predictable. There is a great deal of complicated text to memorize, often in long soliloquies, yet no linear plot and no characterization. This creates an interesting paradox: the production relies on Barrow's compilation of writings on the concept of infinity, including various thought experiments that he explains and "translates" for the lay reader, yet it has no script in the traditional sense of the term. The actors take the ideas from this foundational text and transform them into physical theater that plunges the audience in the ideas as they are enacted more than in the words the actors speak.

Ronconi explains that he wanted a kind of theater that would challenge the audience and the actors alike. "We must leave at home traditional notions of character, story and plot, and the idea of continuous attention from the audience."[11] In *Infinities*, the actor and the audience are in the same boat. "This kind of theater is looking for hypotheses, rather than starting with them. We don't know the final answer. The actors are not above the audience—they often don't understand the text that well, the math, either."[12] Ronconi deliberately puts us in a vulnerable position, as the intermingling of audience and actors makes us feel as visible as the performers. Natalie Crohn Schmitt identifies this approach as one of the defining characteristics of postmodern performance: "The recent interest in designing performance environments that encourage audience members to move about expresses a desire to have the audience witness events from different, changing, and unfamiliar points of view and thus actively, and with self-awareness, participate in the production."[13]

This is consistent with Ronconi's interest in time and temporality in the theater. He jumped on the idea of infinity/infiniteness because he could destroy the apparent coincidence between the time of the play and the time depicted in the play. It's a running show; it's not a whole package that is complete, but an open-ended experience. It avoids the illusion of "complete information" that one gets in traditional theater. Indeed, *Infinities* succeeds not despite but because of its rejection of tradition. Its staging and structure convey its ideas. You are not attending a lecture on mathematics, you are participating in a vivid and imaginative demonstration of stimulating thought experiments. Whether you understand every nuance of these paradoxes is not the point; it is the unmediated immersion in them that matters. What sticks to you once you've emerged from the theater is part of the suspense.

This sounds a good deal like the renowned improvisational director Viola Spolin's approach to performance, which emphasizes game playing rather than acting in the traditional sense, and developing the performance through exercises. "Few of Spolin's exercises are based on conflict, the chief source of intensity in traditional drama, for such exercises, she believes, generate 'emotionalism and verbal battles,' not intuitive responses," writes Schmitt. Like Barrow and Ronconi's *Infinities*, Spolin's "improvisational games focus instead on the immediacy of the actual materials: the game and the actors in present time and space. The 'drama' in this sense becomes the operations performed on stage, operations that serve as opportunities for participants and observers to perceive. The performers' only obligation is to 'share' their games with the audience."[14] This suggests the main difference between text-based science plays like

Copenhagen and *Arcadia* and actor-director-based works like *Mnemonic* and *Infinities*. "When the idea of acting as game-playing is applied to a text, it suggests that the actors and director should play with the text as they might with a prop. A theater game, like one of [John] Cage's 'mesostics,' is intended as a way of experiencing the material rather than of serving the author's intention and representing the text."[15]

To put it another way, the challenge of engaging science on stage is analogous to the dilemma of the translator, who must choose either complete fidelity to the original language at the risk of sounding clumsy, awkward, or leaden in the translation, or the approximate sense of the words in order to achieve elegance of phrasing but possibly sacrificing accuracy. The solution, according to Gillian Beer, is the "transformation" of ideas rather than their translation or explanation.[16] Beer identifies this concept of "transformation" as central to so many works of literature and science, and it is brought to a new performative level in the recent "alternative" science plays. From their methodology to their finished product they embody the notion of transformation that Beer deems essential to the encounter between science and culture. That is, instead of taking the science in isolation and attempting to explain its meaning in lay terms, these works transform it into primarily visual and physical terms.

It is this paradox that makes the "alternative" kind of science play so attractive in skipping the translation of the ideas altogether and going straight to an unmediated use of science on stage. This can be achieved only by radically challenging the conventional way of doing theater. Complicite's entire working methodology and approach to theater is based on two main ideas: change and collaboration. Influenced by teachers like Jacques Lecoq, Philippe Gaulier, and Monika Pagneux, the company is by its own explanation a "constantly evolving ensemble of performers and collaborators. . . . Complicite continuously researches, trains and teaches through workshops. Each workshop is as much a collaborative as an educative process, a sharing and developing of ideas, enabling everybody involved to experiment with and clarify the journey of new work."[17] The program states that "there is no Complicite method. What is essential is collaboration. A collaboration of individuals in order to establish an ensemble, with a common physical and imaginative language, ready to create new work."[18] Complicite workshops source material in collaboration with the actors and writers, in the case of *Mnemonic* drawing on the work of a range of neuroscientists. The final textual product that one can purchase and consult and act from is, to borrow the metaphor used by

McBurney in his biochemistry lecture at the beginning of the show, a map that is open to change and constant adaptation by the user rather than a fixed and permanent script.

Directors' Theater

The shift away from stable, finished text—from text to actor—is quite naturally accompanied by a shift away from author to director. Many contemporary science plays fall into this category. This is similar to Peyret's technique, and to the way in which *Infinities* was developed. The director has a very firm hand, in a sense supplanting the playwright though in no way the sole generator of the work. The directors seem concerned, however, to dispel any notions of autocracy. Interestingly, both Peyret and McBurney publish e-mail excerpts from the actors during the development of the production, as evidence of the collaborative method. Peyret posts these on his Web site for his company Theater Feuilleton 2 (TF2). For his most recent production, *Les Variations Darwin*, the Web site links to a Web documentary-journal of the show's development and rehearsal by "Sainte-Lucie." "In this video-chat room that we have called 'Variations,' Sainte-Lucie receives, questions and collects impressions of those who come through here with their web-cams. These internauts, connected from Shanghai, Boston, Marseille or elsewhere, glance at what is happening onstage, and hang out in the room. Sainte-Lucie will be with us through every performance and invites you into this webroom, very far, very close to the stage in the theater."[19] The idea is to make the process of doing theater as transparent as possible, which, as Natalie Crohn Schmitt reminds us, is "a normal condition of performance, rather than an abnormal condition to be overcome. The ensemble includes the audience; the audience is part of the event, informed of the means of its making."[20] In Peyret's productions, as shown in the film of the rehearsals for *La Génisse et le pythagoricien*, the workshopping and devising of the piece is laid bare. Indeed, it becomes part of the production, validated as much as the performance itself.[21] The Web site for the company contains the scripts as they have evolved for each production, in all their stages (called "partitions"), so that one may consult what text there is for a given piece and see how it was developed. A Peyret and Prochiantz production typically has at least seven or eight *partitions*, all of which are available on their site.

Peyret's insistence on process over neatly presented and packaged ideas relates to what Prochiantz says about the packaging of science. In a way, each feels dissatisfied with his own discipline: Peyret says theater is dead, there are only "theaters," while Prochiantz voices the paradox that the more successful you are in science, the more you have failed. Scientific results, presented as watertight and absolute in respected journals, cover up the *process* of doing science and distort the enterprise of science for the layperson. What is lost is the uncertainty, the moments of hesitation, in scientific investigation. Something similar happens in traditional theater, which hides the process of rehearsal and the theatrical mechanisms that enable a performance (scaffolding, lights, costume production, makeup, scenery). Peyret's science plays uncover this process of theater just as they would uncover the process of doing science, and employ mechanisms like the web to do so.

Like Ronconi, Peyret wants to reach the audience in a more direct way and says that the science plays he creates are *not* about ideas; they are not intellectual. "We do not do scientific theater," write Peyret and Prochiantz in the program to *Les Variations Darwin*; "we in fact do not even know what that means. We are not sure to give you a performance about Darwin, but we know that we did it with him, listening to the music of his thought."[22] They are interested in using science to explore the frontier between human and animal (as in the Darwin trilogy) or, as in Peyret's acclaimed production *Turing Machine*, between human and machine. For example, the second Darwin production, *Des Chimères en automne*, explores the frontier between ape and human by fusing textual source material from Kafka and Darwin; the play synthesizes Kafka's address to the academy by Lucie, the monkey, with selected moments in Darwin's life. However, the show eschews a conventional biographical development. The actors take turns "playing" these characters, shifting constantly in speaking the lines, so that no one actor becomes identified with the part of Lucie or with the part of Darwin. This keeps the audience focused on the ideas in play rather than on a particular story, biography, or character. Events and figures from the history of science (Darwin and evolution) are indeed used, but counterbalanced by fiction (Kafka's Lucie). We do not learn about Darwin and his illness, or the death of his beloved daughter, for the usual dramatic purposes of cathartic immersion in someone else's life; rather, these selected moments in Darwin's life are placed in perspective by being intertwined with glimpses from Lucie's life, intersecting the human and the simian.

Likewise, *Les Variations Darwin*, the third Darwin installment, incorporates aspects of Darwin's life and work into what becomes in the creators' words a "scientific reverie." These are variations: "not an essay, even less a thesis, rather a dream, a fantasy on several Darwinian themes."[23] This play is similar to Barrow and Ronconi's collaboration: what we see is more a series of loosely connected vignettes, episodes, or "thought experiments" than a complete whole with beginning, middle, and end. "Scientific reverie, poetic reverie, another blurry frontier, whose outline necessitates a certain drowsiness. . . . So here's a piece of advice, disguised as instructions: slumber, let yourself dream. . . . We'll try to do the rest."[24] Each vignette or "variation" plunges the audience into a particular experience or idea linked to Darwin, relaying his theories and findings about the brain in relation to emotion, facial expression, and both animal and human behavior. Throughout the play, an actor periodically simulates nausea and staggers offstage, a comical reminder of how viscerally Darwin's ideas affected him. Many of the scenes are particularly interested in gender and sexuality and feature short exchanges between actors appearing to enact ideas that are elaborated in voice-over. In one scene, a mild disagreement between a male and female in what appears to be a domestic situation escalates rapidly into a violent explosion of emotion leading to murder—all stemming from Darwin's observation of how contempt is expressed in the face in his book *The Expression of the Emotions in Man and Animals*.

In one of the play's most striking variations, an overt metaphoric transformation occurs by the suggestive visual connection between a cabbage and a brain. An elegant female actor holds a whole cabbage in her hand, contemplates it, slowly begins to unpeel it and to eat each leaf, then picks up speed and devours half the cabbage in a feeding frenzy, tearing at its innards with her teeth. She then abruptly stops and sits motionless with the cabbage held between her teeth and thus obscuring—indeed replacing, in Magritte-like fashion—her head. A male actor comes over and begins to peel off and eat the leaves of the side of the cabbage that is facing us, then repeats the woman's frenzied devouring of the cabbage; only here he is tearing with his teeth at what appears to be her head, and doing so as if he is passionately kissing her. The distinction between kissing and eating is hilariously called into question.

The voice-overs (often dialogic in form) provide elaborations of the scientific ideas in this "rêve," and such scenes suggest many interpretations to the audience about the interface between animal and human and between violence and tenderness. The play's substantial final episode,

Les Variations Darwin: In a stage image evoking Magritte, an actress sits motionless holding a cabbage between her teeth (the cabbage supplanting her head) while an actor kisses/devours it. © Elisabeth Carecchio.

entitled "Roman" (Novel) takes as its starting point Darwin's complaint later in life that he had lost his earlier passionate engagement with literature, formed by such authors as Milton, Gray, Byron, Wordsworth, Coleridge, and Shelley. All the actors gather in one partial thrust section downstage, each with a carrier bag labeled "The Darwin Project" from which each in turn produces a book to read from. The atmosphere is relaxed and gives the sense of reveling in the prose and the ideas of these authors and how they impinge on Darwin's own work. Reason and pleasure intermingle; the scene reminds us that attenuation of literary pleasure does not necessarily follow from intensified scientific involvement.

Les Variations Darwin has been well received. A review in *Nature* hailed it as "funny, original and anything but didactic. It takes familiar ideas and pursues them through their serious implications *ad absurdam*."[25] Peyret has a great deal of experience in such interdisciplinary work. Before collaborating with Prochiantz, Peyret worked closely with another neuroscientist, Jean-Didier Vincent, on several productions that take an interest in science. Vincent always loved the story of Faust and used to perform the play with puppets in Gerard de Nerval's translation. He convinced Peyret to do something with Faust. A voracious reader, Peyret was keen to engage science; as Vincent puts it, even though "he did not understand the parole of the song, he understood the music."[26] They eventually did a Faust production, about the history, origin, philosophy, and biology of life. The first thing they had to do was find the scenic device that would accommodate their theatrical vision: a membranelike structure bisecting the house to create two audiences. The result was two different plays; one side was called Marguerite, the other Faust. The membrane could open to allow the two parts to communicate; actors could go over and inside it. As Vincent emphasizes, this was highly dramatic, especially when the membrane opened at the end to allow the union of the two parts.

Peyret used this scenography again in La Génisse et le pythagorien (The Heifer and the Pythagorian), the first Prochiantz-Peyret collaboration, and most recently in Les Variations Darwin. In both, the central feature of the staging is a huge movable wall that bisects the stage at an angle to the audience. In the former production, the wall was a combination of solid base and white strips rather like vertical venetian blinds, giving the marvelous feel of solidity and fluidity in one structure. In the latter, the wall has no strips but is covered instead with a scrim so that, depending on the lighting, the wall can seem completely solid or completely diaphanous. In both cases, the wall is partially transparent and mobile, and the actors occasionally penetrate it, either weaving among the strips or walking along the interior of the wall. The lighting enhances the suggestiveness of this visual metaphor, as in several cases actors place their hands on the scrim and their palms appear black against the fabric; paradoxically, they are trapped inside a transparent wall. This visually reinforces the notion of borders that figures so centrally in Peyret and Prochiantz's work.

Likewise, music is used in interesting and original ways in these collaborations. In La Genisse, a mechanical, self-playing piano provides musical interludes, contrasting a pair of live violinists who come and go periodically as part of the action. At times in Les Variations Darwin, electronic notes mimic human tones during the dialogue. As the actors speak, electronic tones mimic both their registers and the duration and quality of their vowels. Machine and human interact; the border between what we ourselves create and what technology can produce is explored and questioned.

Peyret's latest work is a collaboration with Luc Steels on a theatrical piece entitled Le Cas de Sophie K. that investigates the life and personality of Sonya Kowalevski, a nineteenth-century Russian mathematician who died tragically at the age of forty-one. As with his previous explorations of Darwin's life and thought, Peyret's interest in this figure is not primarily biographical; he explains that "her dramatic life will act as a kaleidoscope through which key developments of the era are evoked and interpreted and projected into the present."[27] Peyret and Steels thus use Sonya Kowalevski's life as a means to "explore the nature of mathematical knowledge and mathematical intuition" in ways that challenge the conventional understanding of these arenas as primarily abstract; by contrast, they feel that "mathematical knowledge and intuition appear to be grounded in spatial and bodily imagination."[28] Clearly, Peyret is continuing to find in science and mathematics an enduring source of inspiration for exploring the ways in which audiences can be immersed in a subject that might initially seem foreign and even hostile to them within the world of the stage and its immediate and material possibilities.

What Makes the "Alternative" Science Play Different?

It should be clear by now that many of these nontraditional science plays are linked by a shared concern with essentialism, with "what it is to be human."[29] A review of Complicite's recent *The Elephant Vanishes* reiterates this idea: "The real complexity we take away from [this production], as from earlier Complicite shows . . . , is our reawakened sense of what it is to be human."[30] This is the dominant theme for Peyret as for McBurney, and both want the experience for the audience to be immediate and visceral rather than "intellectual." That leads to a paradox: their plays may seem "intellectually daring,"[31] but they shun intellectualism. They take textual source material only as a starting point: Barrow's notes on different thought experiments relating to infinity; Peyret and Prochiantz's notes on Darwin and Kafka; the books about the Iceman and about neuroscience and memory that inspired *Mnemonic*. The play is constantly developing, building on the input of many collaborators, not written in isolation by one author. The point is not that these plays dispense with textuality but that they *decenter* it; texts become starting points rather than ends in themselves. The idea is also to juxtapose texts, collage-style, and see what new ideas emerge. Inexplicably, this appeared to be very dramatic and intense, according to Jean-Didier Vincent.[32] Putting Kafka and Darwin together, or Darwin and Ovid, or Faust and Turing produced powerful dramatic moments to which general audiences tended to react well, even less educated members, because there are no preconceived notions; they don't know a particular segment comes from Kafka, for example, they just know that it's interesting.

With the text displaced in this way, it is perhaps not surprising that the actor also functions very differently in such pieces. Rather than developing a character, mentally fleshing out the character's story and motives as most actors in traditional theater are trained to do, the actor in the alternative science play is primarily a vessel for ideas. In fact, one might say that mediation does occur but it is through the actor rather than notions of character and plot. The actor must be in effect retrained, weaned off the traditional emphasis on how well the actor becomes the character. Both Ronconi and Peyret frequently double or triple their actors in roles, thus further decreasing the risk of too much absorption and identification on the part of the actor. Peyret inverts this process, in fact, having several actors play one character as well as using doubling. The directors take an almost oppositional approach to their actors, and as Jean-Didier Vincent has put it, Peyret is often very hard on his actors, a "despot," no less.[33] Three weeks before the production, even though there is a great deal of text to memorize, the actors still do not know exactly the final shape and nature of the performance. This is difficult for them, since (as with *Infinities*) there are large chunks of text to memorize, yet not much dialogue to make memorization easier. Also, Peyret borrows elements from previous plays he has done, so that each piece is related in some way; his work is an ongoing story.[34] Audience members who know that he uses this technique make a point of seeing each play, to get this sense of continuity.[35]

Third, a great deal more emphasis is placed on the body in what is intensely physical theater. The actors, who almost have to be gymnasts or acrobats, are constantly in motion. In *Infinities*, actors hang by their feet as they are moved along a track high above the audience's heads in one scene; in another, they run and leap energetically over long rows of books; in a third, they pop out of compartments and move along deep corridors. All the while, they are constantly talking, telling us about Cantor's insanity, the Hotel Infinity, or Borges's Library of Babel. Even the ancient, gray characters in the scene about old age display great physicality: though they are much more languid, sitting under hair dryers or in wheelchairs or lying prone on a table, they are in constant, fluid motion. This is true also of Complicite's actors, who never seem to rest throughout the performance, transforming from one figure to the next and using highly physical, often mime-based techniques to delineate setting and character. The cabbage-devouring scene in *Les Variations Darwin*, described earlier, helps to capture the kind of physicality of the Peyret-Pro-chiantz productions; the play also contains several other stunningly exe-cuted physical feats evocative of animal behavior during mating and courtship rituals.

A fourth feature of the "alternative" science play is the unconventional scenery and the placement of the audience in the house. It is true that many contemporary science plays eschew the realist set. *Copenhagen* and *Molly Sweeney*, for instance, dispense with almost everything but chairs, actors, and spotlights (though the scenery of *Copenhagen* is really much more elaborate than that, involving faux-marble flooring and stalls). However, the suggestion of realism is always there, and through a gestalt-like reflex the audience supplies what is missing—a sofa here, a dining room table there, a door here. In "alternative" science plays like Peyret's and Ronconi's, no such subconscious furnishing occurs because the as-sumption of realism is prevented. In *Infinities*, the audience is moved about in promenade style as well as being directly addressed; and even when seated the audience members are not rigidly separated from the actors, who intermingle with them freely. In *Mnemonic*, audience members are asked to don the eyeshades and finger the leaves that have been placed in plastic bags and taped to each seat in the house, and while doing so to think about their past.

Finally, this movement—if we can call it one—has tended to come (with the exception of Theatre de Complicite) from outside the main-stream British or American theater. Ronconi and Peyret are receiving critical acclaim in their own countries, while their work has yet to receive their English-language premieres. Indeed the plays I am discussing either are not in English or have a strong basis in Continental culture and the-ater practice. *Infinities* is in Italian; Peyret works in French and also incor-porates German, for example in his Faust; neither of these has been trans-lated or performed in England or the States. Complicite is inspired by French experimental theater and the tradition of mime, and the company is multicultural, with actors from Greece, Swiss Germany, France, En-gland, and other countries, all speaking in their own languages and with-out translation or supertitles. Most recently, the company created a cross-cultural performance piece with Japanese actors, *The Elephant Vanishes*, based on the stories of Haruki Murakami, set in Tokyo and exploring the ways in which our increasingly technologized modern culture affects our daily lives. It is interesting how the English reaction to Complicite has pointed up the cultural distance of such collaborations. This highly physi-cal theater group has been received in England as "visual." McBurney notes that "the only people who think of me as 'visual' are the English, and that is because they have an under-developed sense of it, despite a highly developed sense of irony and language. Whereas here [in England] we talk of 'audiences'—listeners—in France attenders at a play are 'les spectateurs'—watchers. I treat the visual with as much respect as the spoken word."[36] Does this perhaps suggest that Anglophone audiences may be more resistant to experimental theater, more likely to expect and demand traditional realism? Certainly the historical evidence supports this suggestion; one thinks, for example, of the hostile reception accorded Ibsen and other progressive playwrights in England at a time of a thriving avant-garde in France.[37]

The Postdramatic Theater

As we have seen, many of these newer, alternative plays share theatrical approaches and methodologies. They also stretch traditional notions of time and space in the theater. "If the moments are significant in themselves, the structure of the work cannot be hierarchical in time, developing to certain moments. Accordingly, the actor's and director's examination of the material tends to become spatial rather than linear, providing a series of extended moments. This interest in the extended moment explains the often-remarked-upon eclecticism of postmodernism: the present moment includes all associations extending in time and space without bounds, without any preconception of what things belong together or of one thing's being subordinate to another."[38] While real-life figures and events do sometimes appear in these alternative plays, they are not there for the sake of biography, and they are offset by fictional characters and events. Darwin, the nineteenth-century mathematician Cantor, the Iceman, and other historical figures are placed in fictional contexts; the real and the imaginary rub up against each other. The difference is in the way the sources are used.

So, we might ask, is it possible to taxonomize, to put a name to the kind of theater I am talking about here? Recent work in theater and performance studies suggests several possible categories that might apply. One is the notion of "devised" theater, which John Schmor defines as theatrical work that is "not initially or primarily scripted by a playwright or dominated by a director-auteur's score, but instead which is created primarily by performers, with designers and directors in intensive collaboration."[39] None of these pieces is really devised theater, because Ronconi and Peyret, for example, have such a firm hand in shaping the final product; the performers are not entirely free to devise either during the process of rehearsing or during the final performances. Another notion is that of "Quantum Theater," David E. R. George's term for a postmodern theater that creates and explores "the space between determinism and uncertainty."[40] George has argued that the old trope of theater as the world (*theatrum mundi*) has become obsolete and even dangerous because it devalues the theater by implication "as a very model of the insincerity, deception, and illusion it locates in everyday life."[41] George protests the "ideological hegemony of the Theatrum Mundi over the theater itself" and suggests in its place a Quantum Theater whose scientific metaphor of uncertainty provides a better model for the way theater and world reflect one another.[42]

Perhaps George's Quantum Theater captures something of what this newer science play represents. However, a more apt and ultimately more productive term might be "postdramatic" theater, as theorized most centrally by Hans-Thies Lehmann in his work *Postdramatisches Theater*. Theater scholar David Barnett points out that one of the key features of postdramatic performances, as opposed to "representational theater," is openness. The starting point of the postdramatic theater is "that there is a theater beyond representation. . . . Actors do not represent characters, text does not represent situation, sets do not represent places. In addition, a postdramatic theater does not structure time."[43] As one can see, for instance, in *Infinities*, "the traditional tensions of the drama retreat to convey the experience of a perpetual present."[44] Time takes on the characteristics of time in dreams as opposed to time in everyday life.[45] This is highly suggestive, especially with regard to Peyret and Prochiantz's use of the term "rêve" to describe the kind of audience experience they seek to foster when they merge science with theater.

On the one hand, "postdramatic performances usually eschew clear coordinates of narrative and character and require therefore considerable effort on the part of the spectator."[46] On the other hand, the popularity of plays like *Infinities* indicates either that audiences don't mind the effort or that such effort is simply not required. What is significant is that text, actor, and audience interact in wholly different ways from their status in conventional theater. Language may be liberated from "interpretive limitation," but, Barnett cautions, "we should not necessarily understand the status of the text as free but free-floating, hovering above a vast range of contexts and meanings, *each with multiple points of contact with life and the experience of the audience*."[47]

So far I have been discussing the opposition—though I prefer to think of it as a complementarity—between the more conventional, representational science play and the alternative, "postdramatic" ones. But there is a wider context still for this discussion. The "postdramatic" science play relates directly to—and in fact may be emerging out of—the ongoing debates about what is "performance" and what is "theater." Over the last twenty years or so, the subject of performance has received a tremendous amount of attention and has permeated several other fields in addition to establishing itself as a subject of study in its own right, "performance studies." What this generally implies is that theater is old-fashioned, conservative, and not innovative or groundbreaking, while performance studies implies cutting-edge advances in theatrical endeavors. Many theater scholars find this "false dichotomy" troubling, especially if it is indicative of future directions in theater studies. "Just as the popular concepts of *theatricality* and *performativity* are in danger of meaning everything, and thus nothing, so too is our discipline in danger of crossing so many borders that it loses its way. . . . Does our discipline have an identity?"[48]

A similar polarization has occurred between the kind of theater we would call realist and the postmodern art of performance. Vanden Heuvel notes that realism has become, by a kind of "negative consensus," synonymous with "conciliation, assimilation, adaptation, and resignation" to the dominant, hegemonic, "bourgeois, patriarchal, oppressive, and oedipal" discourses that it replicates.[49] Thus, both "theater" and "realism" are often dismissed as out of date and indeed insidious. As Vanden Heuvel points out, "Realist plays are nowadays almost automatically associated with an 'old-fashioned' kind of text, one that is suspect aesthetically as well as politically." However, these newer, postdramatic science plays represent a challenge to such dichotomies.

David Knight maintains that "if education is a matter of lighting fires rather than filling bottles, then we should still see professing as a performance art."[50] With regard to science in the theater, this reinforces the argument that it is not about educating audiences with didactic plays full of science, but exposing them in various ways to science, whether directly or implicitly. There are also other kinds of science-performance that defy easy categorization. A collaborative "theater-as-experiment" entitled *Perceptions and Realities: The Science of Theater* was premiered at the London International Festival of Theater in June 2004. The director Katie Mitchell, playwright Caryl Churchill, and neuropsychologist Mark Lythgoe of the Institute of Child Health collaborated to produce this piece whose starting point is the question: "How does the mind relate to the world and how can theater help to investigate it?"[51] The piece is basically a highly entertaining multimedia lecture by Lythgoe that involves the audience in interactive exercises. It has several sections, rather like Peyret and Prochiantz's collaborations. One segment is, in Lythgoe's words, "an attempt to create a performance that is so abstract that it forces the audience to let go and take in all the information at face value without any of the 'top-down perceptual processing' that superimposes forms of prior knowledge on what we experience." Another section involves actors attempting to represent "ambiguity" by seemingly incongruous actions, for example, combining a "happy voice" with a "disgusted body."[52] What is being tested is "the strength of the audience's likely desire to resolve such ambiguities by creating an illusory explanation (such as 'mania')."[53] Rather like a thought experiment, the piece works on the premises of theater itself to "embody, dramatise and ponder these profound perceptual puzzles in an interactive exchange with the public."[54] Many other examples of this kind of science-performance-theater can be found, such as Anna Furse's *Yerma's Eggs* (about infertility and the techniques and psychological implications of assisted reproductive technology, with a literary inspiration by Lorca) and the collaborative medical-musical performance piece *Viewing the Instruments* discussed earlier.[55]

Furse's performance piece calls for "projected biological material" such as a four-dimensional ultrasound image of a fetus in utero, images of the egg cell and the sperm fertilizing it, and images of ICSI (intra-cytoplasmic sperm injection). Echoing very much the sentiments of other contemporary practitioners who use science in their productions, like Peyret and Ronconi, Furse writes in her program notes to *Yerma's Eggs*: "I didn't want to write a play, impose my authority on a single-track narrative. . . . I wanted to get under the skin of the subject via the body in performance—expressionistically, viscerally, and reflect the complexity and contradiction via a layering of elements."[56]

Will such science-performance pieces be the dominant trend in the future of science on stage, or will the postdramatic works that have been the focus here prove most lasting? Or will we see some combination of the two? Whatever the future direction of science in the theater, it is certain that the interaction of science and the stage has produced some powerful works that cover an astonishing range of scientific subjects and theatrical approaches. Such diversity is to be welcomed as a sign of a healthy and thriving genre, one that actively reflects and engages key aspects of the scientific, medical, and theatrical cultures of our times. Science on stage has clearly moved beyond the two dominant paradigms of the genre, represented first by the figure of Faust and then by that of Dr. Strangelove and his cold war ilk. The recent examples explored here look beyond these two types to embrace a multiplicity of themes and theatricalities. Besides making innovative and provocative theater, they can foster greater social awareness of science. All of these qualities were demonstrated by the opera *Dr. Atomic*, a new work by composer John Adams and director Peter Sellars that premiered in 2005 in San Francisco and drew acclaim for its musical and dramatic treatment of J. Robert Oppenheimer and his work at Los Alamos. Above all, recent science plays defy C. P. Snow's pessimistic forecast of a widening rift between the two cultures and instead encourage each culture to learn about the other through the interactive and persistently experimental medium of performance.

It remains to be said that as with any play, a good science drama must entertain as a piece of theater—it must work on the stage. The plays chosen for investigation in this study represent the best of what the intersection of science and theater has to offer in this respect first and foremost, as well as in their textual engagement with scientific ideas. Most of all, these plays offer a continuing, collective rumination on what Gillian Beer calls "the place of science in culture."[57] They demonstrate that theater can play a vital role in helping us understand our encounter with the increasingly urgent questions and issues posed by science.

Appendix

● ● ● ● ● ● ● ● ● ● ● ● ●

Four Centuries of Science Plays: An Annotated List

I have compiled this list with the generous help of Harry Lustig. It is an ongoing project to which we are constantly adding new entries; this appendix represents our most up-to-date version of the list as it now stands, and it is a much-expanded version of the one we published in *American Scientist*: see Harry Lustig and Kirsten Shepherd-Barr, "Science as Theater," *American Scientist* 90 (November–December 2002), 550–55. Another version of this list, compiled by Shepherd-Barr, Lustig, and Brian Schwartz, will be made available on the Web. Please note that an asterisk at the end of an entry indicates sponsorship by the Alfred P. Sloan Foundation and, in many cases, production in the foundation's First Light Festival of new plays dealing with science.

ASTRONOMY/ASTROPHYSICS/COSMOLOGY

Andreyev, Leonid. *To the Stars*. (1907) An astronomer who lives apart from society is blind to the changing world around him, even as family members try to make him aware of what is happening.

Godfrey, Paul. *The Blue Ball*. (1995) About the space program. Commissioned by the National Theatre, England.

Gunderson, Lauren. *Background*. (2003) The physicists Ralph Alpher and Robert Herman deduced the existence of cosmic background radiation in 1948. A few observers looked for it, and it was found accidentally in 1965. Alpher's life and lack of recognition and the history of cosmology are recounted going backward in time, illuminating both.

Hunter, Maureen. *Transit of Venus*. (1992) In France at a time when society was rapidly expanding its knowledge of the earth and the cosmos, an ambitious astronomer and the women who love him exemplify the conflicting needs of men and women. First Canadian play ever staged by the Royal Shakespeare Company.

Landau, Tina. *Space*. (2001) A meditation on astrophysics; a New Age play about a professor of neuropsychiatry and part-time therapist and his three patients who claim to have been abducted by space aliens.

Lavik, Erin. *Galileo Walking among the Stars.* (2004) Written by an assistant professor of biomedical engineering at Yale, the play imagines Galileo, Kepler, Harriot, and Gene Kelly building a spaceship in order to be among the stars rather than just look at them from afar.

Miyagawa, Chiori, and James Lattis. *Comet Hunter.* (2003) Explores the life of eighteenth-century German astronomer Caroline Herschel, the first woman to identify a comet, and her brother William, the famous astronomer and identifier of the planet Uranus.*

BIOLOGICAL SCIENCES

Mac Low, Clarinda, James Hannaham, and Tanya Barfield. *The Division of Memory.* (2001) At the end of his life, an African American research biologist reflects on his place in the twentieth century.

Nachtmann, Rita. *Thread of Life.* (2003) The role of Rosalind Franklin in the discovery of the structure of DNA.

Theatre de Complicite. *Mnemonic.* (2000) About memory, connection, chaos, and evolution, with the Iceman as its starting point. Opens with a "lecture" on the biochemistry of memory.

CHEMISTRY

Djerassi, Carl, and Roald Hoffman. *Oxygen.* (2000) With scenes alternating between contemporary Sweden and eighteenth-century France and England, the play asks, who should be awarded the first, fictional, "Retro-Nobel" award for a scientific discovery before the twentieth century?

Gorki, Maxim. *Children of the Sun.* (1905; trans. Stephen Mulrine, 1999) About an idealistic chemist who wants to be left alone, is uninterested in the realities around him and unsympathetic to the claim that science should serve society.

Harrison, Tony. *Square Rounds.* (1992) The story of Fritz Haber, the chemist whose discoveries led to extreme good and extreme evil—the development of fertilizer, which in turn would alleviate hunger, and the invention of the gas that was used in World War I on a devastating scale.

Horovitz, Israel. *Promises.com.* (2003) Set in the world of research chemistry, this drama explores questions of love, integrity, promises, and compromise.

Mamet, David. *The Water Engine.* (1977) According to Mel Gussow, an inventor "manages to remove the H from H_2O and—eureka!—he invents an engine that uses plain distilled water as fuel" (Gussow review in *Theatre on the Edge*, 199–201).

Poliakoff, Stephen. *Blinded by the Sun*. (1996) How the media affect modern scientific research, with a prominent team of scientists pressured to fake a groundbreaking discovery.

Thiessen, Vern. *Einstein's Gift*. (2003) The Canadian playwright's award-winning drama about chemist Fritz Haber.

DARWIN AND EVOLUTIONARY THEORY

Lawrence, Jerome, and Robert E. Lee. *Inherit the Wind*. (1955) A courtroom drama-documentary play about the Scopes trial pitting Darwin's theory of evolution against the Bible.

Peyret, Jean-François, and Alain Prochiantz. *Des Chimères en automne, ou l'impromptu de Chaillot*. (2003) A three-hander contemplating the differences—and similarities—between humans and apes, with Darwin and Kafka as its motivating and interconnected forces.

Peyret, Jean- François, and Alain Prochiantz. *Les Variations Darwin*. (2004) Further meditations on the frontier between human and ape, with Darwin's ideas and Ovid's *Metamorphoses* as inspirational sources.

Sherman, Jonathan Marc. *Evolution*. (2002) According to the *New Yorker*, (30 September 2002) this is "a morality play ... about an academic studying Charles Darwin who is offered a job in the entertainment industry." Directed by Lizzie Gottlieb.

Wertenbaker, Timberlake. *After Darwin*. (1998) Set in alternating scenes on the *Beagle* as Captain FitzRoy and Darwin clash, and the actors portraying them in the modern play-within-the-play reflect the increasing tension between the two men.

Whittell, Crispin. *Darwin in Malibu*. (2003) "Darwin has wound up in a beach house overlooking the Pacific with a girl young enough to be his daughter. Believing that the heated debate about the *Origin of Species* is far behind him, Darwin now finds guidance from cheap tabloid horoscopes and trashy beach reading. But when his old friend Thomas Huxley washes up on the beach with the bishop of Oxford he finds himself entangled in a life and death comedy about God, science, love, loss and the sex life of barnacles." Blurb on Methuen Web site.

Wilson, Snoo. *Darwin's Flood*. (1994) Darwin, Nietzsche, Jesus, and Mary Magdalene meet and match wits.

GENERAL SCIENCE/SCIENTISTS

Berger, Glen. *Great Men of Science, nos. 21 and 22*. (1998) Set in Paris 1793–94, the Reign of Terror, it examines the ideals of the Enlightenment scientist in the context of political and social upheaval. Directed at the Yale Dramat by Wier Harman, with music by John Kline.

Brecht, Bertolt. *Life of Galileo.* (1939, 1947) Scenes focusing on key moments in the scientist's life and work as he struggled with the church.

Büchner, Georg. *Woyzeck.* (1837) The subject of a bizarre scientific experiment becomes unbalanced and murders his lover in a jealous frenzy.

Clyman, Bob. *The Secret Order.* (1999–2000) About the pressures threatening to destroy a young scientist.*

Djerassi, Carl. *Calculus (Newton's Whores).* (2003) A revisionist look at Newton and his "whores" in the Royal Society.

Egan, David. *The Fly-Bottle.* (2003) A philosophy graduate of Harvard, Egan pits three famous philosophers, two also working within the discipline of mathematics, against each other: Wittgenstein, Popper, and Russell.

Goethe, Johann Wolfgang von. *Faust.* (1808, part I; 1831, part II) A scientist and scholar has grown weary of his learning and, aided by a powerful accomplice, regains his youth and pursues pleasure, with mixed consequences.

Golding, William. *The Brass Butterfly.* (1958) Set in third-century Rome, the play features a scientist far ahead of his time who invents an explosive missile, a steamship, the pressure cooker, and other dangerous technology; but the wise emperor rejects these innovations and suggests that the scientist devote himself to gardening.

Hampton, Christopher. *The Talking Cure.* (2002) About the relationship between Freud and Jung.

Jonson, Ben. *The Alchemist.* (1610) Gullible customers line up to benefit from the promises of alchemy purveyed by three skilled con men.

MacLeish, Archibald. *Heracles.* (1965) The protagonist, a scientist at the zenith of his career, has just been awarded a Nobel Prize. The adulation pleases him, but he is also aware of the futility and the costs of contemporary scientific discovery.

Marlowe, Christopher. *Doctor Faustus.* (1604 and 1616 versions, known as "A" and "B" texts) Archetypal science play dramatizing the relationship of knowledge to power and the dangers of overreaching scientific exploration.

Moving Being Productions, Wales. *Alternating Currents.* (2004) The life and legacy of Nikola Tesla.

Ravenhill, Mark. *Faust (Faust Is Dead).* (1997) Famous philosopher arrives in Los Angeles and is lauded as a star. On a TV show he announces the Death of Men and the End of History and then meets a young man on the run from his father, a software tycoon. Doollee Web site.

Scriblerus [John Gay]. *Three Hours after Marriage.* (1717) This comedy features a doctor-scientist, Fossil, who has just married a much younger and very clever woman and is already finding it too much to cope with as the curtain opens. The climactic scene is set in Fossil's amazing museum, which houses his scientific collection, including an alligator that mysteriously comes to life.

Shadwell, Thomas. *The Virtuoso.* (1676) Thinly veiled lampoon of the scientist Robert Hooke and other Royal Society members.

Shaw, Jane Catherine. *The Lone Runner: The Mythical Life Journey of Nikola Tesla.* (1999) A puppet version of the story of Nikola Tesla.

Stanley, Jeff. *Tesla's Letters.* (1999) Ensemble Studio/First Light production.* Another play about "the underappreciated Nikola Tesla, whose contribution to science is with us every time we flick on a light switch. Even though the invention of electricity is associated in most people's minds with Thomas Edison, it was Tesla's discovery of the principle of the rotating field which is the basis of most alternating-current technology and which truly ushered in the age of electrical power." Elyse Sommer, curtainup.com review.

Stoppard, Tom. *Galileo.* (1970) A screenplay that challenges Brecht's *Galileo* and was originally intended for performance in the London Planetarium; manuscript is in the Harry Ransom Humanities Research Center, University of Texas at Austin. It has been published in a special issue of the journal *Areté* (Spring–Summer 2003).

Wertenbaker, Timberlake. *Galileo's Daughter.* (2004) Maria Celeste, Galileo's illegitimate daughter, lives in a convent in Florence. Based on the book by Dava Sobel and directed by Peter Hall for the Theatre Royal Bath.

Wilson, Lanford. *The Mound Builders.* (1986) Explores the world of archaeology and how it relates to contemporary life.

GENETICS, CLONING, AND REPRODUCTIVE TECHNOLOGY

Burns, Elizabeth. *Autodestruct: The Ultimate Cure for Cancer.* (2001) A scientist clones his way to immortality, but at what price? Performed at Jesus College, Oxford.

Churchill, Caryl. *A Number.* (2002) A father confronts three of his adult sons, two of whom are clones of the first. Churchill uses the scientific possibility of cloning to address the basic human question of where personality comes from—nature or nurture? Royal Court Theatre, 2002.

Djerassi, Carl. *An Immaculate Misconception.* (2001) As the *New Yorker* puts it, a play by "the inventor of the birth control bill, about sex in the age of fertility treatments."

Furse, Anna. *Yerma's Eggs.* (2003) Performance piece exploring infertility, with textual elements by Spanish playwright Lorca.

Glaspell, Susan. *The Verge.* (1921) A female botanist creates new plants in all-consuming experiments that bring her into conflict with her family.

McGrath, Tom. *Safe Delivery.* (1999) Gene therapy and laboratory politics seen through the eyes of a young female researcher with a secret. Cowritten with the author's daughter, a geneticist, and funded by the Wellcome Trust as part of its Science on Stage and Screen program.

Stephenson, Shelagh. *An Experiment with an Air-Pump.* (1998) Set on the thresholds of two new centuries, the play focuses on medical experimentation's ethical dimensions and the idea of scientific progress.

Mathematics

Auburn, David. *Proof.* (2000) A young, insecure female mathematics student, and not her psychotic mathematical genius of a father, turns out to have solved a fiendishly difficult theorem.* (Made into a film starring Gwyneth Paltrow and Anthony Hopkins)

Barrow, John. *Infinities.* (2002) Five scenarios of different mathematical "thought experiments" on the concept of infinity. Presented in Milan at the Teatro Piccolo, directed by Luca Ronconi, in spring 2002 and spring 2003, and in Valencia, Spain, in 2002 (sponsored by the Sigma Tau Foundation, Italy).

Brenton, Howard. *The Genius.* (1983) Brash young mathematician discovers how to make the most powerful atomic weapon and has to live with this knowledge and its consequences.

Doxiadis, Apostolos. *Incompleteness: A Play and a Theorem.* (2004) The final days of Kurt Gödel in the hospital when he is refusing food and is challenged by a nutritionist who attempts to apply Gödel's own teachings on logic to persuade him to eat; he is also visited by the vision of mathematician David Hilbert.

Groff, Rinne. *The Five Hysterical Girls Theorem.* (2000) A group of mathematicians meet at a British seaside resort in 1911 for a caper about number theory.

Hauptmann, Ira. *Partition.* (2003) A fantastical drama about math genius S. Ramanujan.

Poliakoff, Stephen. *Breaking the Code.* (1986) A play about code-breaker Alan Turing and "Enigma" during World War II.

Stoppard, Tom. *Arcadia.* (1993) Chaos theory, mathematics, landscape gardening, and literary history collide in a country house in England, alternating contemporary with eighteenth-century scenes.

Wellman, Mac. *Hypatia, or the Divine Algebra*. (2000). "Hypatia, the 5th century mathematician, pagan philosopher and inventor, was considered so inherently dangerous that Christian monks found it necessary to drag her through the streets of Alexandria, Egypt, before dismembering and then burning her body. The play follows Hypatia's imaginary trajectory from that spectacle through 8th century Byzantium and then on to the early 20th century." Elyse Sommer, curtainsup.com review.

MEDICINE/DOCTORS

Bernstein, Elsa. *Dämmerung: Schauspiel in fünf Akten* [*Twilight: A Drama in Five Acts*]. (date?) In this naturalist play, a female eye surgeon successfully treats the daughter of a man who is against educated women and wins his love. Poised to sacrifice medicine for marriage, she "finds misery instead." *MLA Newsletter*, Winter 2003.

Brieux, Eugene. *Les Avariés*. (1909) Controversial play dealing with syphilis and its personal and public consequences.

Brook, Peter, and Marie-Hélène Estienne. *The Man Who*. (2002) Based on the Oliver Sacks neurology case study "The Man Who Mistook His Wife for a Hat," this "theatrical research" was first performed, in English, in Zurich in 1994.

Dunn, Nell. *Cancer Tales*. (2001) Verbatim stories of five women with cancer (performed at King's College, London, in May 2003).

Edson, Margaret. *Wit*. (1999) Set in a hospital ward, the play depicts an uncompromising professor of metaphysical poetry who endures grueling treatments for ovarian cancer and is sustained by her love of Donne's Holy Sonnets and the late-budding friendships she finally allows herself to have. Adapted for an HBO film starring Emma Thompson, Christopher Lloyd, and Eileen Atkins, directed by Mike Nichols.

Flanagan Davis, Hallie. *Spirochete*. (1938) The epic history of syphilis from its origins to the present day, told as a Living Newspaper and performed by the Federal Theatre Project.

Flood, David H., and Rhonda L. Soricelli. *The Seventh Chair: An Audience Encounter*. (1993) Brings together several famous fictional doctors, such as Ibsen's Dr. Stockmann, in a therapy session for "doctors who are experiencing a crisis in professional identity." Published in a special issue of the journal *Literature and Medicine*.

Friel, Brian. *Molly Sweeney*. (1994) Based on neurologist Oliver Sacks's short story "To See and Not See," the play deplicts a blind woman given an operation and the surprising and painful consequences of gaining sight after so many years.

Ibsen, Henrik. *An Enemy of the People*. (1882) Doctor discovers dangerous bacteria in town spa waters, but instead of appreciation he meets townspeople's wrath as politics trumps science.

A MNESIA CURIOSA
CHEKHOV LIZARDBRAIN

Kingsley, Sidney. *Men in White.* (1933) Prototypical *ER* episode that depicts a hospital and doctors treating patients, acclaimed for its pioneering use of realism in staging surgeons at work.

Kushner, Tony. *Angels in America.* (1991) Although not thought of as first and foremost a "science play," this was one of the first plays to deal in depth with the science and medical treatment of AIDS.

Saul, Oscar, and H. R. Hays. *Medicine Show.* (1940) A Living Newspaper revealing the inner workings of hospitals to Mr. Average Citizen. Performed on Broadway. Unpublished, in the archives of the Federal Theatre Project.

Shaw, George Bernard. *The Doctor's Dilemma.* (1906) Which of two patients should receive the experimental cure for tuberculosis: the worthy but plodding country doctor, or the amoral but gifted artist?

Smith, Anna Deavere. *Untitled.* (2000) The renowned author of *Fires in the Mirror* and *Twilight: Los Angeles* portrays all parts in this one-woman show about doctors, patients and their narratives. Yale Tercentennial performance event at Yale Medical School.

Musicals and Operas

Ausbury, Steven, and Anthony Burr. *Biospheria.* (2001) An "environmental opera" based on aspects of the unsuccessful "environmentalist experiment" at Biosphere 2, a multi-million-dollar greenhouse erected north of Tuscon in the early 1990s.

Drama of Works. *The Ballad of Phineas P. Gage.* (2002) In 1848 an iron rod passed through the head of Phineas P. Gage, and he survived. But how did he live? The award-winning puppet theater explores through puppetry, music, and poetry this groundbreaking neurological case.*

Glass, Philip, Arnold Weinstein, and Mary Zimmerman. *Galileo Galilei.* (2002) An opera that dramatizes the life of "the inventor of modern science," with an emphasis on the fun of doing scientific research.

Lessner, Joanne Sydney, and Joshua Rosenbaum. *Einstein's Dreams.* (2002) Original musical adaptation of Alan Lightman's novel *Einstein's Dreams.* Produced by Brian Schwartz.

Lessner, Joanne Sydney, and Joshua Rosenbaum. *Fermat's Last Tango.* (2000) A musical comedy in which a mathematician (based on the real-life figure of Andrew Wiles) is led to the "Aftermath" by his nemesis, Fermat, to meet Pythagoras, Newton, Euclid, and Gauss, who assess his proof of Fermat's last theorem.

McKee, Anne Gaud, and Christian Denisart. *The Oracle of Delphi.* (2000) With choreography by Markus Schmid. A pantomime about P.A.M. Dirac's discovery of the positron and antimatter.

SCHRÖDINGER'S GIRLFRIEND

Reale, Robert, and Willie Reale. *Quark Victory.* (2000) A musical in which a young girl journeys through a subatomic world occupied by dancing electrons and singing neutrinos.

Riordan, Arthur, and Bell Helicopter. *Improbable Frequency.* (2005) A musical comedy set in Dublin during World War II and interweaving espionage, politics, and the figure of Schrödinger, who was resident in Ireland during this time. Staged at the Abbey Theatre by the Rough Magic Theatre company.

Skipitares, Theodora. *Defenders of the Code.* (1987) A musical of scientific facts that covers everything from creation myths to theories of eugenics, and incorporates Plato's *Republic*, Darwin's *Origin of Species*, and Watson's *Double Helix* into a collage.

Tapper, Albert M., and James Racheff. *Imperfect Chemistry.* (2000) A musical comedy in which a pair of geneticists seek a cure for baldness.

Wilson, John F., and Grace Hawthorne. *The Electric Sunshine Man.* (1978) A musical about Thomas Edison, geared for young children.

Wilson, Robert, and Philip Glass. *Einstein on the Beach.* (1976) Groundbreaking postmodern opera. American premiere at the Metropolitan Opera House.

Zimet, Paul, and Ellen Meadow. *Star Messengers.* (2001) A quasi opera that shows Galileo, Tycho Brahe, and Kepler interacting with three harlequins from one of Galileo's books.

See also curtainup.com Web site containing an annotated list of science- and math-related plays (http://www.curtainup.com/scienceplays.html - 20k - 2004–04–23).

PHYSICS

See Charles A. Carpenter, *Dramatists and the Bomb*, for a more complete record.

Congdon, Constance. *No Mercy.* (1994) About the first atomic bomb test and the men involved in the nuclear program.

D'Andrea, Paul, and Jon Klein. *The Einstein Project.* (2000) "The Einstein met here was an intense, charismatic but not very loving or lovable man. His relationship with his overly sensitive son [is] used to most tellingly [illustrate] the contradictions and flaws in his personality. He had an intense relationship with the boy, yet overchallenged him cruelly and ends up putting his public concerns before the boy's needs." Elyse Sommer, curtainup.com review.

Davis, Hallie Flanagan. $E = mc^2$. (1948) A Living Newspaper showing what nuclear energy is and outlining its dangers and possibilities for humankind.

Dürrenmatt, Friedrich. *The Physicists.* (1962) Tragicomedy about a physicist who has discovered a theory of everything but hides himself away in an asylum to prevent its dissemination, little knowing that his fellow inmates, Einstein and Newton, are in fact spies, and the director of the institution a mad dictator in disguise.

Frayn, Michael. *Copenhagen.* (1998, 2000) In the afterlife, Niels Bohr, Werner Heisenberg, and Margrethe Bohr attempt to get at the truth of what happened during Heisenberg's famous visit to Bohr in Copenhagen in 1941.

Friedman, Robert Marc. *Remembering Miss Meitner.* (2002) In the chamber of historical memory, the codiscoverer and explicator of nuclear fission confronts Otto Hahn, who could have helped her to receive a share of his Nobel Prize, and Manne Siegbahn, who, while providing a refuge for her in his Swedish laboratory, did not provide her with any wherewithal for continuing her research.

Giron, Arthur. *Moving Bodies.* (2000) Dramatizes the biography and contributions of the great, idiosyncratic physicist Richard Feynman, including his role in the building of the atomic bomb and the explanation of the explosion of the space shuttle *Challenger.**

Hoar, Stuart. *Rutherford.* (2000) New play about Ernest Rutherford, the great New Zealand physicist. "The play focuses on the enigma that was Rutherford, follows his obsession with science and probes his personal relationships with his wife Mary, daughter Eileen, and friend and colleague, the Russian, Kapitza." http://www.nzedge.com/heroes/rutherford.html.

Jahnn, Hans Henny. *Trümmer des Gewissens (Der staubige Regenbogen)/The Ruins of Conscience: The Dusty Rainbow.* (1959) A physicist sees too late that his research may have terrible consequences, and commits suicide because he is unable to avert the threat of mass destruction.

Johnson, Terry. *Insignificance.* (1983) Imagines a meeting between Albert Einstein, Marilyn Monroe, and Joe DiMaggio in a hotel room as Einstein is being pursued by a senator from the House Un-American Activities Committee.

Kipphardt, Heinar. *In the Matter of J. Robert Oppenheimer.* (1964; trans. Ruth Speirs and performed at Mark Taper Forum 1967–68) Documentary theater based on the 1954 security clearance hearings by the Atomic Energy Commission.

MacColl, Ewan. *Uranium 235.* (1952) A dynamic atomic energy play by Theatre Workshop enacting the history of the atom from its birth to the present time.

Martin, Steve. *Picasso at the Lapin Agile.* (1996) A farcical comedy that imagines a meeting between Picasso and Einstein in a café in Paris, with a surprise appearance by Elvis.

Morgan, Charles. *The Burning Glass.* (1953) One of many post-Hiroshima plays in which the scientist is an individual who poses a grave threat to humanity. Here, a weather control machine would allow the sun's radiation to be concentrated on any specific spot on earth.

Parnell, Peter. *QED*. (2001) Alan Alda portrays Richard Feynman, the Nobel Prize–winning physicist, in a one-man show that is part physics lesson, part biography.

Penniston, Penny. *Now Then Again*. (2000) Set in Fermilab National Particle Accelerator Lab in Batavia, Illinois. The play's complicated nonlinear structure evokes the physics that the characters discuss.*

Pinner, David. *Newton's Hooke*. (2003) A play based on the lives and politics of those key figures of the Royal Society, Newton and Hooke, who helped to shape the institution as well as the science of physics.

Simms, Willard. *Einstein, a Stage Portrait*. (1996?) A one-man show about Albert Einstein, starring John Crowther. American Jewish Theatre.

Sinclair, Upton. *A Giant's Strength*. (1948) A play about atomic physicist Barry Harding.

Stoppard, Tom. *Hapgood*. (1988) Spy games interwoven with Heisenberg's uncertainty principle.

Wells, Matthew. *Schrödinger's Girlfriend*. (2002) The eponymous author of the "Schrödinger's Cat" paradox applies the lessons of quantum mechanics to a torrid love affair.*

Wilson, Lanford. *Rain Dance*. (2003) Set in Los Alamos, New Mexico, 1945, on the eve of the birth of the atomic bomb. In the tranquil beauty of the desert, four individuals involved in the historic project count down to its inevitable conclusion. As the culmination of their work approaches, each wrestles with the weight of responsibility for an event that will change the world forever.

RADIATION

Fenwick, Jean-Noël. *Les Palmes de M. Schutz*. (1992) A French play about Marie and Pierre Curie's discovery of radium, set in a realistic stage laboratory. A great stage success, it became a film and was also translated into Italian as *Amore e chimica*.

Frontczak, Susan Marie. *Manya*. (2002) One-woman show about Marie Curie.

Mabou Mines. *Dead End Kids*. (1980) According to Natalie Crohn Schmitt, the production is "about the discovery and uses of radiation . . . [and] examined our fascination with transformation, beginning with our effort to turn base metal into gold and ending with our success in turning atoms into bombs." Natalie Crohn Schmitt, *Actors and Onlookers*, 122–23.

Mullin, Paul. *Louis Slotin Sonata*. (2001) Events surrounding the accidental death of a scientist at Los Alamos from radiation exposure.*

STRING THEORY

Jones, Charlotte. *Humble Boy*. (2002) In this contemporary version of *Hamlet*, a neurotic astrophysicist in a dysfunctional family tries to create a "theory of everything" out of string theory and general relativity.

Reingold, Jacquelyn. *String Fever*. (2003) A forty-year-old violinist turned teacher deals with her midlife crisis through string theory and the attentions of the attractive physicist who tells her about it.*

Speier, Susanna. *Calabi Yau*. (2002) A "string-theory comedy" in which New York subway workers try to build a particle accelerator in abandoned subway tunnels.

TECHNOLOGY

Čapek, Karel. *R.U.R.* (1921) "Rossum's Universal Robots" examines the devastating effect of robots on society.

Kopit, Arthur. *Y2K*. (1999) Deals with the threats to our privacy when computer hackers invade our lives via the Internet.

Reich, Steve, and Beryl Korot. *Three Tales*. (2002) A multimedia collaboration, consisting of three sequences for live instrumentalists, singers, and video projection, amounting to "a parable of man's Faustian bargain with technology." M. Billington in the *Guardian*, 19 September 2002.

Theatre de Complicite/Haruki Murakami. *The Elephant Vanishes*. (2003; revival 2004) Set in Japan, a series of vignettes exploring the dehumanizing effect of technology on people's lives. Opens with a prologue about the speed of light, Einstein, and the second law of thermodynamics.

Rozin, Seth. Reinventing Eden.

Notes

·······

INTRODUCTION

1. Carol Rocamora, "Scientific Dramaturgy," *Nation* (5 June 2000), 50.
2. Susan Stanford Friedman, "Definitional Excursions: The Meanings of Modern/Modernity/Modernism," *Modernism/Modernity* 8, no. 3 (September 2001), 501.
3. For the sake of clarity and brevity, throughout the book I use the term "science" broadly to encompass the natural sciences, physics, mathematics, medicine, and technology.
4. Judy Kupferman, "Science in Theater," *PhysicaPlus: Online Magazine of the Israel Physical Society* (2003), 1.
5. John Stokes, "Guilt Edged" (a review of Shelagh Stephenson's *Mappa Mundi*), *Times Literary Supplement* (22 November 2002), 19.
6. David E. R. George, "Quantum Theater—Potential Theater: A New Paradigm," *New Theater Quarterly* 5, no. 18 (May 1989), 172–73.
7. Michael Blakemore, "From Physics to Metaphysics and the Bomb," *New York Times* (9 April 2000), 20.
8. I have written about this merging of form and content in articles in *Interdisciplinary Science Reviews*, *Gramma*, and *American Scientist*, all listed in the bibliography. See also comments by Mark Berninger on the way science playwrights are "explicitly constructing a connection between the form of the dramatic texts and the various paradigm changes in scientific theory" in Berninger, "A Crucible of Two Cultures: Timberlake Wertenbaker's *After Darwin* and Science in Recent British Drama," *Gramma* 10 (2002), 109.
9. C. P. Snow, *The Two Cultures and a Second Look* (Cambridge: Cambridge University Press, 1984), 16.
10. Physicists Harry Lustig and John Wheeler, for example, in Harry Lustig and Kirsten Shepherd-Barr, "Science as Theater," *American Scientist* 90 (November–December 2002), 550–55.
11. Jane Goodall, *Performance and Evolution in the Age of Darwin: Out of the Natural Order* (London: Routledge, 2002), 1.
12. See Kirsten Shepherd-Barr, "Acting Out the Search for Infinity," *Physics World* (July 2003), 38.
13. Peter Langdal, *Copenhagen* seminar, Niels Bohr Institute (September 1999); see Niels Bohr Institute Web site for proceedings.
14. See Patrick D. Murphy, ed., *Staging the Impossible: The Fantastic Mode in Modern Drama* (Westport, Conn.: Greenwood Press, 1992).
15. See, for example, Nick Ruddick, "The Search for a Quantum Ethics: Michael Frayn's *Copenhagen* and Other Recent British Science Plays," *Journal of the Fantastic in the Arts* 11, no. 4 (2001), 415–31; Victoria Stewart, "A Theater of Uncertainties: Science and History in Michael Frayn's 'Copenhagen,'" *New Theater Quarterly* 15, no. 60 (November 1999), 301–7; Lustig and Shepherd-Barr, "Sci-

ence as Theater," 550–55; Kirsten Shepherd-Barr, "*Copenhagen*, Infinity, and Beyond: Science Meets Literature on the Stage," *Interdisciplinary Science Reviews* 28, no. 3 (September 2003), 193–99; Shepherd-Barr, "*Copenhagen* and Beyond: The 'Rich and Mentally Nourishing' Interplay of Science and Theater," *Gramma* 10 (2002), 171–82; Karen C. Blansfield, "Atom and Eve: The Mating of Science and Humanism," *South Atlantic Review* 68, no. 4 (Fall 2003), 1–16; *Interdisciplinary Science Reviews* 27, no. 3 (September 2002), 161–247, special issue on science and theater. Further references are provided in the bibliography.

16. Snow, *Two Cultures*, 25.
17. Rocamora, "Scientific Dramaturgy," 49.
18. Baz Kershaw, *The Radical in Performance: Between Brecht and Baudrillard* (New York: Routledge, 1999), 5.
19. Quoted in Peter Rodgers, "Novel Approaches to Science," *Physics World* (November 2002), 38.
20. Snow, *Two Cultures*, 16.
21. National Public Radio's *Talk of the Nation: Science Friday* shows on science and theater aired 20 October 2000 (featuring David Auburn's *Proof*), 22 February 2002 (on the physics play *QED*), and 23 January 2004 (discussing *Fermat's Last Tango*).
22. For example, symposia on *Copenhagen's* use of history, science, and theatricality have been held in New York; Washington, D.C.; Raleigh, North Carolina; Princeton; Cambridge, Massachusetts; and Albuquerque, New Mexico. Outside the United States there have been symposia on the play in Copenhagen; Cambridge, England; and Germany.
23. Kirsten Shepherd-Barr, *Ibsen and Early Modernist Theatre, 1890–1900* (Westport, Conn.: Greenwood Press, 1997), 12–13, 74.
24. Martin Esslin, " 'Dead! And Never Called Me Mother!': The Missing Dimension in American Drama," *Studies in the Literary Imagination* (Fall 1988), 23, 27.
25. Gillian Beer, *Open Fields: Science in Cultural Encounter* (Oxford: Oxford University Press, 1996), 321.
26. Ibid., 322.
27. Tom Stoppard, *Arcadia* (London: Faber and Faber, 1993), 62.
28. Glynne Wickham, "Drama in a World of Science," in *Drama in a World of Science and Three Other Lectures* (London: Routledge and Kegan Paul, 1962), 52–53.
29. Ibid., 47.
30. Ibid., 53.
31. Ibid., 56.

CHAPTER ONE: THE TRADTION OF SCIENCE PLAYS

1. William Shakespeare, *Coriolanus*, III.i, in *William Shakespeare: The Complete Works*, ed. Alfred Harbage (New York: Viking Penguin, 1969).
2. Ibid., I.iv.30–35
3. William Shakespeare, *As You Like It*, II.vii, ll. 139–43, in *William Shakespeare: The Complete Works*, ed. Alfred Harbage (New York: Viking Penguin, 1969), 257.

4. William Shakespeare, *Henry V*, prologue, ll. 3–33, in *William Shakespeare: The Complete Works*, ed. Alfred Harbage (New York: Viking Penguin, 1969).

5. George, "Quantum Theater," 171. George points to two valuable sources for the history of the *theatrum mundi* metaphor: Ernst Robert Curtius, *European Literature and the Latin Middle Ages* (Princeton, N.J.: Princeton University Press, 1953), and Jonas A. Barish, *The Anti-theatrical Prejudice* (Berkeley: University of California Press, 1981).

6. In fact, this use of simple props to make the stage double as the cosmos follows the opening moments of the play, in which Andrea plays with a "big astronomical model" with metal rings, balls, and bands representing earth, sun, moon, and stars. It is Galileo's dismissal of this useless gift from the Court of Naples that prompts him to show his pupil the movement of the sun using more down-to-earth props. This shift from the small-scale inanimate model that is only looked at to the participatory, human-scale model that is interactive subtly marks the kind of transformation Brecht's epic theater strives toward; it also enacts the idea of the stage as cosmos.

7. Berninger, "Crucible of Two Cultures," 107.

8. Ibid.

9. Richard Allen Cave, *English Dramatists: Ben Jonson* (New York: St. Martin's Press, 1991), 78.

10. Peter Porter, review of *Doctor Faustus* at Young Vic, *Times Literary Supplement*, (16 March 2002).

11. Christopher Marlowe, *Doctor Faustus*, ed. David Bevington and Eric Rasmussen (Manchester: Manchester University Press, 1993), introduction, 34.

12. Porter, review of *Doctor Faustus*.

13. Ibid.

14. For an in-depth analysis of the use of astronomy in the play, and Marlowe's understanding of contemporary science, see the introduction to the Revels edition of the play, 27–29.

15. Marlowe, *Doctor Faustus*, introduction, 35.

16. Brecht wrote three versions of the play; see Eric Bentley's introduction and appendix in Bertolt Brecht, *Galileo*, trans. Charles Laughton (New York: Grove Press, 1966).

17. Beer, *Open Fields*, 331.

18. Ben Jonson, *The Alchemist*, ed. Simon Trussler (London: Nick Hern Books, 1996), 52–53, II.v, ll. 20–33.

19. Roger Sales, *Christopher Marlowe* (New York: St. Martin's Press, 1991), 109–10.

20. Anne Barton, "Catiline," in *Critical Essays on Ben Jonson*, ed. Robert N. Watson (New York: G. K. Hall, 1997), 43.

21. Cave, *English Dramatists*, 77.

22. Ibid., 77–78.

23. Ibid., 5.

24. For analyses of *The Virtuoso*, see, for instance, Roslynn D. Haynes, *From Faust to Strangelove: Representations of the Scientist in Western Literature* (Baltimore: Johns Hopkins University Press, 1994); and M. A. Orthofer, "The Scientist on the Stage: A Survey," *Interdisciplinary Science Reviews* 27, no. 3 (2002), 173–83.

25. John Gay, *Three Hours after Marriage: A Comedy, as It Is Acted at the Theatre Royal* (London: Bernard Lintot, 1717), 10.

26. Eric Bentley, "The Science Fiction of Bertolt Brecht," introduction to Brecht, *Galileo*, 12.

27. Brecht quoted in ibid., 16.

28. Ibid., 18. Bentley distinguishes the two versions still further: "*Galileo I* is a 'liberal' defense of freedom against tyranny, while *Galileo II* is a Marxist defense of a social conception of science against the 'liberal' view that truth is an end in itself." Ibid., 18–19n.

29. Ibid.

30. James K. Lyon, *Bertolt Brecht in America* (Princeton, N.J.: Princeton University Press, 1980), 178.

31. Haynes, *From Faust to Strangelove*, 280.

32. Howard Brenton, *Plays: 2* (London: Methuen, 1989), xi.

33. Mel Gussow, *Theater on the Edge: New Visions, New Voices* (New York: Applause, 1998), xii.

34. I am indebted to Lewis Wolpert for sharing with me the text of his playlet "Good Evening, Galileo."

35. Stoppard quoted in John Fleming, *Stoppard's Theater: Finding Order amid Chaos* (Austin: University of Texas Press, 2001), 67.

36. Ibid., 68. Stoppard's text has recently been published in *Areté* 11 (Spring/Summer 2003), along with an introduction by the playwright. A thorough summary and analysis of Stoppard's *Galileo* can be found in Fleming, *Stoppard's Theater*, 66–81, including substantial excerpts from the text.

37. Wolpert, "Good Evening, Galileo."

38. Thomas S. Kuhn, *The Structure of Scientific Revolutions*, 2nd ed. (Chicago: University of Chicago Press, 1970).

39. Brecht, *Galileo*, 85.

40. Haynes, *From Faust to Strangelove*, 291.

41. Ibid.

42. Ibid.

43. Thornton Wilder, *The Skin of Our Teeth*, in *Three Plays* (New York: Avon Books, 1976), 122.

44. Stoppard, *Galileo*, in *Areté*.

45. Fleming, *Stoppard's Theater*, 66.

46. Stoppard, "A Note on *Galileo*," in *Areté*, 5–6.

47. Fleming, *Stoppard's Theater*, 67.

48. Ibid., 68.

49. Stoppard, *Galileo*, quoted in Fleming, *Stoppard's Theater*, 270n9.

50. Fleming, *Stoppard's Theater*, 73.

51. Ibid., 269n4.

52. Friedrich Dürrenmatt, *The Physicists*, trans. James Kirkup (New York: Grove Press, 1964).

53. Annabel Patterson in Frank Lentricchia and Thomas McLaughlin, eds., *Critical Terms for Literary Study* (Chicago: University of Chicago Press, 1995), 138.

54. Ibid.; see also J. Hillis Miller, *Speech Acts in Literature* (Stanford, Calif.: Stanford University Press, 2001).

55. Alisa Solomon, *Re-dressing the Canon: Essays on Theater and Gender* (New York: Routledge, 1997), 3.
56. Brecht, *Galileo*, 100.
57. Miller, *Speech Acts in Literature*, 19.
58. The show is available on DVD or video from the Clay Mathematics Institute, which helped to sponsor its development and production.
59. Mary G. Winkler, "The Anatomical Theater," *Literature and Medicine* 12, no. 1 (Spring 1993), 67.
60. Ibid., 66.
61. David Knight, "Scientific Lectures: A History of Performance," *Interdisciplinary Science Reviews* 27, no. 3 (2002): 217.
62. Ibid..
63. Winkler, "Anatomical Theater," 69.
64. Ibid.
65. An invaluable source on this subject is Joseph Roach, *The Player's Passion: Studies in the Science of Acting* (Newark: University of Delaware Press, 1985).
66. For excellent analyses of plays concerned specifically with moral issues relating to scientific pursuit, see Allen E. Hye, *The Moral Dilemma of the Scientist in Modern Drama* (Lewiston, N.Y.: Edwin Mellen Press, 1996).

CHAPTER TWO: WHY THEATER? THE APPEAL OF SCIENCE PLAYS NOW

1. Ruddick, "Search for a Quantum Ethics," 416.
2. Ibid., 424.
3. John L. DiGaetani, *A Search for a Postmodern Theater: Interviews with Contemporary Playwrights* (Westport, Conn.: Greenwood Press, 1991), 268.
4. Harold Pinter, "Writing for the Theater," reprinted in *Modern Drama: Plays, Criticism, Theory*, ed. W. B. Worthen (New York: Harcourt Brace Jovanovich, 1994), 598–99.
5. Ibid., 599.
6. Natalie Crohn Schmitt, *Actors and Onlookers: Theater and Twentieth-Century Scientific Views of Nature* (Evanston, Ill.: Northwestern University Press, 1990), 127.
7. Ruddick, "Search for a Quantum Ethics," 415.
8. Jim Burge, "In Search of Science on the Big Screen," *Interdisciplinary Science Reviews* 27, no. 3 (Autumn 2002): 168.
9. Frayn at *Copenhagen* seminar, Niels Bohr Institute, Copenhagen, Denmark (19 November 1999).
10. Snow, *Two Cultures*, 16, 71.
11. See, for example, Michel Serres, *Le Tiers-instruit* (Paris: Gallimard, 1991), published in English as *The Troubadour of Knowledge*, trans. Sheila Faria Gloser and William Paulson (Ann Arbor: University of Michigan Press, 1997); and various other works by Serres that are especially relevant to this study, including *A History of Scientific Thought: Elements of a History of Science* (Oxford: Blackwell, 1995); *Conversations on Science, Culture, and Time*, with Bruno Latour, trans. Roxanne Lapidus (Ann Arbor: University of Michigan Press, 1995); and *The Birth of Physics*, trans. Jack Hawkes (Manchester: Clinamen Press, 2001).

12. Ruddick, "Search for a Quantum Ethics," 428.
13. Dennis Overbye, "Lab Coat Chic: The Arts Embrace Science," *New York Times* (28 January 2003), F1.
14. Murray Smith, "Darwin and the Directors: Film, Emotion and the Face in the Age of Evolution," *Times Literary Supplement* (7 February 2003), 13.
15. Michael Frayn quoted in Anthony Gardner, "The Writer and the Appliance of Science," *Independent* (8 May 2001), 1. He reiterates this idea in a radio interview with Andrew Marr about *Democracy*, his next play after *Copenhagen* (BBC Radio 4, 7 July 2003). In response to Marr's suggestion that people go to the theater to think, Frayn says that instead they do so "to be involved." The distinction is an important one, implying that the appeal of plays like *Copenhagen* is not first and foremost intellectual in nature.
16. Ruddick, "Search for a Quantum Ethics," 428.
17. Ibid., 427.
18. See Susan Bennett, *Theater Audiences* (New York: Routledge, 1997).
19. Glynne Wickham, "Drama in a World of Science," in *Drama in a World of Science and Three Other Lectures* (London: Routledge and Kegan Paul, 1962), 47.
20. Ibid., 52, 51.
21. Beer, *Open Fields*, 331. See also Bruno Latour and Steve Woolgar, *Laboratory Life: The Construction of Scientific Facts* (Princeton, N.J.: Princeton University Press, 1979), and Latour, *Science in Action* (Cambridge, Mass.: Harvard University Press, 1988).
22. Michael Frayn quoted in Gardner, "The Writer and the Appliance of Science," 1.
23. John Maddox, "We Should Be Told," *Times Literary Supplement* (25 April 2003), 4.
24. Author interview with Alain Prochiantz, Paris (22 December 2003).
25. Alan Lightman, "Art That Transfigures Science," *New York Times* (15 March 2003), B9.
26. Ibid.
27. Lawson Taitte, "Stage Sees a Scientific Breakthrough," *Dallas Morning News* (27 October 2000).
28. Peter Brook, *Threads of Time: A Memoir* (London: Methuen, 1998), 220.
29. Ibid., 221.
30. Roger A. Crockett, *Understanding Friedrich Dürrenmatt* (Columbia: University of South Carolina Press, 1998), 105.
31. Frayn quoted in Gardner, "The Writer and the Appliance of Science," 1.
32. Rodgers, "Novel Approaches to Science," 38.
33. Ibid.
34. Ciara Muldoon and Peter Rodgers, "A Brief History of Art and Science," *Physics World* (November 2002), 40.
35. Frayn quoted in Gardner, "The Writer and the Appliance of Science," 1.
36. Sarah Lyall, "Enter Farce and Erudition," *New York Times* (25 October 1999), E1.
37. Michael Frayn quoted in Claire Armitstead, "'Write the Same Thing Over and Over,'" *Guardian* (31 January 2002).
38. Murray B. Peppard, *Friedrich Dürrenmatt* (New York: Twayne, 1969), 72.

39. Timberlake Wertenbaker quoted in DiGaetani, *Search for a Postmodern Theater*, 268.

40. Ibid.

41. See, for example, such popular science books as *Great Feuds in Science: Ten of the Liveliest Disputes Ever*, *Great Feuds in Medicine: Ten of the Liveliest Disputes Ever*, and *Great Feuds in Technology: Ten of the Liveliest Disputes Ever*, all by Hal Hellman and published by Wiley in 1999, 2002, and 2004, respectively.

42. The European Network of Science Communication Teachers (ENSCOT) has recently made available a resource pack based entirely around "Science Controversies: Learning by Doing." Like Hellman's books, this emphasizes the drama inherent in such conflicts as the Mesmer controversy, the Pasteur-Pouchet controversy, Ramon y Cajal and Golgi, and other scientist clashes, as well as cold fusion, the "gay gene," ESP, homeopathy, and phrenology. I am indebted to Neil Hook of the University of Glamorgan, Wales, for bringing this to my attention.

43. I am indebted to Robert Gross for these suggestions about the emergence of science plays.

44. See, for example, Stephen Wilson, *Information Arts: Intersections of Art, Science, and Technology* (Cambridge, Mass.: MIT Press, 2002), especially chapter 8, for a comprehensive survey of contemporary artists' use of science and technology. Wilson leaves theater out of his book; in fact, on page 10, where he states what the book does *not* consider ("the popular media of science fiction, literature, cinema, and television"), the theater is not even mentioned.

45. Ruddick, "Search for a Quantum Ethics," 417.

46. Ibid., passim.

47. See, for example, Andrew Ross, ed., *Science Wars* (Durham, N.C.: Duke University Press, 1996); Sandra Harding, *Is Science Multicultural? Post-colonialism, Feminism, and Epistemologies* (Bloomington: Indiana University Press, 1998), Harding, *Whose Science? Whose Knowledge? Thinking from Women's Lives* (Ithaca, N.Y.: Cornell University Press, 1991); Harding, *The Science Question in Feminism* (Ithaca, N.Y.: Cornell University Press, 1986); and Reed Dasenbrock, *Truth and Consequences: Intentions, Conventions, and the New Thematics* (Pittsburgh: Pennsylvania State University Press, 2000), which revives the notion of objective truth.

48. Einstein appears in Steve Martin's *Picasso at the Lapin Agile*; Dürrenmatt's *The Physicists*; Terry Johnson's *Insignificance*; *Einstein: A Stage Portrait* by Willard Simms; Jerry Meyers's one-man show "You Know Al; He's a Funny Guy" (1981); and several operas. Darwin figures in *After Darwin* by Timberlake Wertenbaker; *Evolution* by Jonathan Marc Sherman; *Darwin in Malibu* by Crispin Whittell; and Snoo Wilson's *Darwin's Flood*. See appendix for further examples.

49. Heinar Kipphardt, *In the Matter of J. Robert Oppenheimer*, trans. Ruth Speirs (New York: Hill and Wang, 1968), 44.

50. Wertenbaker quoted in DiGaetani, *Search for a Postmodern Theater*, 273.

51. Allen E. Hye, *The Moral Dilemma of the Scientist in Modern Drama* (Lewiston, N.Y.: Edwin Mellen Press, 1996), 5.

52. Overbye, "Lab Coat Chic."

53. Peter Langdal, *Copenhagen* seminar proceedings (19 November 1999), on Niels Bohr Institute's Web site.

54. Michael Frayn, *Copenhagen* (London: Methuen, 2003), 63–64.
55. Ibid., p. 66.
56. Stoppard, *Arcadia*, 38.
57. Ibid., 47–48.
58. Ibid., 63.
59. Ibid., 75.
60. Ibid.
61. Frayn, *Copenhagen*, 87.
62. Beer, *Open Fields*, 330.
63. Ibid., 323.
64. Tony Harrison, *Square Rounds* (London: Faber and Faber, 1992), 5. See also Vern Thiessen's award-winning play *Einstein's Gift*, which is also about Haber.
65. Beer, *Open Fields*, 331.

CHAPTER THREE: "LIVING NEWSPAPERS" AND OTHER PLAYS
ABOUT PHYSICS AND PHYSICISTS

1. Dürrenmatt, *The Physicists*, 39.
2. Ruddick, "Search for a Quantum Ethics," passim.
3. George, "Quantum Theatre," 172.
4. Charles A. Carpenter, *Dramatists and the Bomb: American and British Playwrights Confront the Nuclear Age, 1945–1964* (Westport, Conn.: Greenwood Press, 1999).
5. Lustig and Shepherd-Barr, "Science as Theater," 550–55.
6. Another cold war play that has to do with nuclear holocaust is *Where Are Flowers?* by Lorraine Hansberry; this unfinished play imagines a postnuclear society, but then so does (one can argue) Beckett's *Endgame*.
7. Haynes, *From Faust to Strangelove*, 281.
8. Ibid. Sinclair's play is "an earnest, if somewhat clumsy, attempt at naturalistic thesis drama," with "concrete knowledge and ideas about the nuclear situation" and a plea for "world-government control of atomic power." Carpenter, *Dramatists and the Bomb*, 37.
9. For a concise and penetrating analysis of postmodernism vis-à-vis history and historiography, see Patrick Joyce, "A Quiet Victory," *Times Literary Supplement* (26 October 2001), 15.
10. Attilio Favorini, "Representation and Reality: The Case of Documentary Theater," *Theater Survey* 35, no. 2 (November 1994), 33.
11. Harold Clurman, *The Collected Works of Harold Clurman: Six Decades of Commentary on Theater, Dance, Music, Film, Arts, and Letters*, ed. Marjorie Roggia and Glenn Young (New York: Applause Books, 1994), 426. Note that the archives of the FTP are beginning to be downloaded and made accessible on the Library of Congress Web site project "American Memory," an invaluable resource that includes scripts and promptbooks, correspondence, original charters, and other documents relating to the founding and operation of the FTP.
12. Ibid.
13. Hallie Flanagan Davis, introduction to *Federal Theater Plays*, ed. Pierre de Rohan (New York: Random House, 1938), vii–viii.

14. Ibid., ix–x.
15. In this sense *Spirochete* harks back to Brieux's notorious 1913 play about the disease (discussed in chapter 7).
16. Joanne Bentley, *Hallie Flanagan: A Life in the American Theater* (New York: Alfred A. Knopf, 1988), 386.
17. Review in Smith College newspaper, no date available.
18. Ibid.
19. Douglas McDermott, "The Living Newspaper as a Dramatic Form," *Modern Drama* 8, no. 1 (1965), 92.
20. Janelle Reinelt, "Performance and the Documentary Impulse" (unpublished paper for ASTR seminar on documentary theater, 2001), 2. This is part of a larger project under the title *Public Performances: Race and Nation in the Theater of our Time*, forthcoming from University of Michigan Press.
21. Ibid., 3–4.
22. Ibid., 4.
23. Ibid., 5.
24. Ibid.
25. Ibid., 6.
26. Carpenter, *Dramatists and the Bomb*, 35.
27. Ibid., 35–36.
28. Ibid., 36.
29. Kipphardt, *In the Matter of J. Robert Oppenheimer*, 16.
30. Clurman, *Collected Works*, 426.
31. McDermott, "Living Newspaper as a Dramatic Form," 92.
32. Clurman, *Collected Works*, 427.
33. McDermott, "Living Newspaper as a Dramatic Form," 92.
34. MacColl, *Journeyman*, 245.
35. Ibid.
36. McDermott notes that "perhaps the surest index of English feeling [of high praise for the Living Newspapers] was the series of Living Newspapers staged by the Unity Theater in imitation of the Federal Theater." McDermott, "Living Newspaper as a Dramatic Form," 94. He also refers to an article about the English Living Newspapers by John Collier, "Theater Documentary," *Theater Arts* 31 (July 1947), 65–66.
37. Favorini, "Representation and Reality," 36.
38. McDermott, "Living Newspaper as a Dramatic Form," 85.
39. Ewan MacColl, *Uranium 235*, in Howard Goorney and Ewan MacColl, *Agit-Prop to Theatre Workshop: Political Playscripts 1930–1950* (Manchester: Manchester University Press, 1986), 109.
40. MacColl, *Journeyman*, 117.
41. Ibid., 252.
42. MacColl, *Uranium 235*, 111–14.
43. Critics cannot seem to agree on what mode or genre the play represents; see, for example, Peppard, *Friedrich Dürrenmatt*, 68–69 ("comedy," "grotesque," "serious elements," "buffoonery," "cabaret effects," "grotesque comedy"); and Clurman, who notes the play's combination of "comedy melodrama" and "detective story" (*Collected Works of Harold Clurman*, 521).

44. MacColl, *Uranium 235*, 76.
45. Ibid., 125.
46. Ibid., 126.
47. Ibid., 129.
48. Ibid.
49. MacColl, *Journeyman*, 251.
50. Ibid.
51. Ibid., 251–52.
52. Ibid., 263.
53. Ibid.
54. Haynes, *From Faust to Strangelove*, 288.
55. Kipphardt, *In the Matter of J. Robert Oppenheimer*, 16.
56. Ibid., 26–27.
57. See, for example, books by Allen Hye and Roslynn D. Haynes quoted in this study (especially Haynes, *From Faust to Strangelove*, 285–88).
58. Kipphardt, *In the Matter of J. Robert Oppenheimer*, 25.
59. Ibid., 22–23.
60. Ibid., 65.
61. Ibid., 45.
62. Ibid., 48.
63. Ibid., 71.
64. Ibid., 79.
65. Ibid., 63.
66. Dürrenmatt, *The Physicists*, 75.
67. Ibid., 76.
68. Ibid.
69. Ibid., 78–79.
70. Ibid., 80.
71. Ibid.
72. Ibid., 80–81.
73. Ibid., 82.
74. Ibid., 89.
75. Stoppard, *Arcadia*, 4–5.
76. Shelagh Stephenson, *An Experiment with an Air-Pump* (London: Methuen, 1998).
77. Dürrenmatt quoted in Crockett, *Understanding Friedrich Dürrenmatt*, 108.
78. Roslynn Haynes comes the closest when she mentions the Möbius strip as a metaphor for Möbius's dilemma: "caught in a paradoxical situation where action and nonaction amount to the same thing, just as the Möbius loop has only one surface and one edge." But she does not connect the metaphoric function of this symbol to its performative potential (Haynes, *From Faust to Strangelove*, 283).
79. Dürrenmatt, *The Physicists*, 84.
80. Stoppard, *Arcadia*, 74.
81. Marlowe, *Doctor Faustus*, I.i.65, p. 114.
82. Ibid., I.i.51, p. 113.
83. Brian Greene, *The Elegant Universe: Superstrings, Hidden Dimensions, and the Quest for the Ultimate Theory* (London: Vintage, 2000), 15.

84. John Maddox, "We Should Be Told," *Times Literary Supplement* (25 April 2003), 3–4 (a review of new books by Richard Dawkins and Peter Atkins).

85. Ibid.

86. Rob Ritchie, introduction, *Terry Johnson, Plays: I* (London: Methuen, 1993), ix.

87. Brenton, *The Genius*, 169.

88. Howard Brenton, *Plays: 2* (London: Methuen, 1989), xii.

89. Ibid., xii–xiii.

90. Brenton, *The Genius*, 196.

91. Ibid., 210.

92. Brenton, *The Genius*, 191.

93. Ibid., 192.

94. Ibid., 194.

95. Ibid., 194–95.

96. Ibid., 195.

97. Terry Johnson, *Insignificance*, in *Plays: I* (London: Methuen, 1993), 14.

98. Ibid., 16.

99. Ibid., 19.

100. Ibid., 50.

101. Ibid., 59.

102. Ibid., 61.

103. Fleming, *Stoppard's Theater*, 176.

104. Ibid.

105. Ibid., 179.

106. Ibid., 181.

107. Ibid.

108. Ibid., 180; quoting Kelly, *Cambridge Companion to Tom Stoppard*.

109. Schmitt, *Actors and Onlookers*, 118.

CHAPTER FOUR: COPENHAGEN INTERPRETATIONS:
THE EPISTEMOLOGY OF INTENTION

1. Frayn quoted in Gardner, "The Writer and the Appliance of Science," 1.

2. Michael Frayn in conversation with the author, Jesus College Colloquium on *Copenhagen* (24 November 2002).

3. Frayn, *Copenhagen*, 53.

4. Kirsten Shepherd-Barr and Gordon M. Shepherd, "Madeleines and Neuromodernism: Reassessing Mechanisms of Autobiographical Memory in Proust," *A/B: Autobiography Studies* 13, no. 1 (Spring 1998), 39–60.

5. Frayn, *Copenhagen*, 35.

6. Peter Langdal, *Copenhagen* seminar (19 November 1999), Niels Bohr Institute Web site proceedings.

7. John Lahr, "Bombs and Qualms," review of *Copenhagen* in the *New Yorker* (24 April and 1 May 2000), 219. In conversations held at the Cambridge, England, colloquium on *Copenhagen*, Frayn indicated that Lahr had misquoted him; what he had actually told Lahr was that he hoped that the play would get produced, but that if it were not, he would reluctantly allow it to be done as a radio play.

8. Jeremy Bernstein, e-mail correspondence (29 November 2002).

  9. Stewart, "Theatre of Uncertainties," 302.
 10. Frayn, *Copenhagen*, 67–68.
 11. Ibid., 71–72.
 12. Ibid., 98.
 13. Ibid.
 14. Ibid., 98–99; emphasis added.
 15. Ibid., 24.
 16. Ibid., 58–59.
 17. Ibid., 59–60.
 18. Ibid., 66–67.
 19. Ibid., 68–69.
 20. Ibid., 69.
 21. Ibid.
 22. Ibid., 69–70.
 23. Ibid., 52.
 24. Ibid., 71.
 25. Ibid., 72.
 26. Ibid.
 27. Ibid.
 28. Ibid., 73.
 29. Ibid., 77–78.
 30. Ibid., 92.
 31. Frayn in conversation with the author at the Cambridge colloquium on *Copenhagen* (23 November 2002).
 32. Frayn, *Copenhagen*, 91.
 33. Quoted in John Cornwell, *Hitler's Scientists: Science, War, and the Devil's Pact* (New York: Viking, 2003), 411.
 34. Frayn, *Copenhagen*, 86–87.
 35. Shepherd-Barr and Shepherd, "Madeleines and Neuromodernism," 44–45.
 36. Blakemore, "From Physics to Metaphysics," 7, 20.
 37. Ibid, 20.
 38. Thomas Powers, "A Letter from Copenhagen," *New York Review of Books* (14 August 2003), 55.
 39. Ibid., 56.
 40. Ibid., 55.
 41. Michael Frayn, " 'Copenhagen' Revisited," *New York Review of Books* (28 March 2002), p. 23. This has become known as the "post-postscript" to the play; it is reprinted in the 2003 edition of the play.
 42. Klaus Gottstein, "Heisenberg Visit to Copenhagen and the Bohr Letters," at SEPP Web site, www.sepp.org/NewSEPP/Heisenberg-Bohr.htm.
 43. Langdal, *Copenhagen* seminar.
 44. Ibid.
 45. Frayn, " 'Copenhagen' Revisited," 21–22.
 46. James Glanz, "Frayn Takes Stock of Bohr Revelations," *New York Times* (9 February 2002).

47. Rainer Karlsch, *Hitlers Bombe* (Munich: DVA, 2005). I am grateful to Jeremy Bernstein for clarifying some of the findings and implications of Karlsch's book.
48. Frayn, " 'Copenhagen' Revisited," 24.

CHAPTER FIVE: EVOLUTION IN PERFORMANCE:
THE NATURAL SCIENCES ON STAGE

1. Jean-François Peyret and Alain Prochiantz, program notes to *Les Variations Darwin*, Théâtre National de Chaillot, salle Gémier (November–December 2004); translation by Lisbeth Shepherd.
2. See, for example, Nancy Bunge, "The Social Realism of *Our Town*: A Study in Misunderstanding," in *Thornton Wilder: New Essays*, ed. Martin Blank, Dalma Hunyadi Brunauer, and David Garrett Izzo (West Cornwall, Conn.: Locust Hill Press, 1999), 349–64.
3. Goodall, *Performance and Evolution in the Age of Darwin*.
4. Wertenbaker, *After Darwin*, 27. For an understanding of how the play developed from its original version, undergoing significant changes in the process, see, for example, Sara Freeman, "Adaptation after Darwin: Timberlake Wertenbaker's Evolving Texts," *Modern Drama* 45, no. 4 (2002), 646–62.
5. Wertenbaker, *After Darwin*, 57.
6. Benedict Nightingale, review of *After Darwin*, *New York Times* (9 August 1998), 4.
7. Wertenbaker, *After Darwin*, 66.
8. For further critical discussion of Darwin and FitzRoy's encounters with the Fuegians, see several chapters in Beer, *Open Fields*.
9. Ibid., 32.
10. Ibid., 43.
11. The discussion that follows is based on this production, recorded for the National Video Archives on 21 August 1998.
12. Ruddick, "Search for a Quantum Ethics," 427–28. Mark Berninger also devotes detailed analysis to the title of the play in "Crucible of Two Cultures," 115–17.
13. Wertenbaker, *After Darwin*, 65.
14. Ibid.
15. Ibid., 70.
16. Peter Morton, *The Vital Science: Biology and the Literary Imagination, 1860–1900* (London: Allen and Unwin, 1984), 4.
17. Ibid.
18. Ibid.
19. Ibid., 4–5.
20. Wertenbaker, *After Darwin*, 56.
21. Ibid., 423.
22. "Theatre and the Laboratory: Do They Have Glass Ceilings?" Royal Society (20 March 2003). Wertenbaker was a guest speaker on this panel, which was chaired by Brenda Maddox, considering the roles of women in science and in the theater.
23. Ibid.
24. Nightingale, review of *After Darwin*.

25. Ibid.
26. Berninger, "Crucible of Two Cultures," 113.
27. Ruddick, "Search for a Quantum Ethics," 423.
28. Sylvie Coyaud, review of John Barrow's *Infinities*, *Interdisciplinary Science Reviews* 27, no. 3 (Autumn 2002), 246.
29. Stephenson, *Experiment with an Air-Pump*, 70–71.
30. Ibid., 31–32.
31. Ibid., 68.
32. Ibid., 52.
33. Ibid., 71.
34. Ibid., 89.
35. Ibid., 88.
36. Michael Billington, "A Number," *Guardian* (27 September 2002), 23.
37. Marek Kohn, review of Matt Ridley's book *Nature via Nurture: Genes, Experience and What Makes Us Human*, *Independent on Sunday* (23 March 2003), 1.
38. Ibid., 2.
39. Ibid.
40. Ibid.
41. Caryl Churchill, *A Number* (London: Nick Hern Books, 2002), 16–17.

CHAPTER SIX: MATHEMATICS AND THERMODYNAMICS IN THE THEATER

1. Marston Morse, quoted in Stanley Gudder, *A Mathematical Journey* (New York: McGraw-Hill, 1976).
2. David Auburn, *Proof* (New York: Faber and Faber, 2001), 15–16.
3. Blansfield, "Atom and Eve," 7.
4. Auburn, *Proof*, 9–10.
5. Ibid., 15.
6. Ibid., 19.
7. Ibid., 47.
8. Ibid.
9. Ibid., 79.
10. David Auburn on National Public Radio, *Talk of the Nation: Science Friday* (20 October 2000).
11. Paul Edwards, "Science in *Hapgood* and *Arcadia*," in *The Cambridge Companion to Tom Stoppard*, ed. Katherine E. Kelly (Cambridge: Cambridge University Press, 2001), 179.
12. Of the many articles written about *Arcadia*, see, for example, Heinz Antor, "The Arts, the Sciences and the Making of Meaning: Tom Stoppard's *Arcadia* as a Post-structuralist Play," *Anglia* 116, no. 3 (1998), 326–54; Therese Fischer-Seidel, "Chaos Theory, Landscape Gardening, and Tom Stoppard's Dramatology of Coincidence in *Arcadia*," in *Emerging Structures in Interdisciplinary Perspective*, ed. Rudi Keller and Karl Menges (Tübingen: Francke, 1997); Fleming, *Stoppard's Theater*; D. Jernigan, "Tom Stoppard and 'Post-Modern Science': Normalizing Radical Epistemologies in *Hapgood* and *Arcadia*," *Comparative Drama* 37 (2003), 3–35; Edwards, "Science in *Hapgood* and *Arcadia*"; Prapassaree Kramer and Jeffrey Kramer, "Stoppard's *Arcadia*: Research, Time, Loss," *Modern Drama* 40, no.

1 (Spring 1997), 1–10; Lucy Melbourne, " 'Plotting the Apple of Knowledge': Tom Stoppard's *Arcadia* as Iterated Theatrical Algorithm," *Modern Drama* 41, no. 4 (Winter 1998), 557–72; Susanne Vees-Gulani, "Hidden Order in the 'Stoppard Set': Chaos Theory in the Content and Structure of Tom Stoppard's *Arcadia*," *Modern Drama* 42, no. 3 (Fall 1999), 411–26.

13. Melbourne, " 'Plotting the Apple of Knowledge,' " 565.

14. Marvin Carlson, *Theories of the Theatre: A Historical and Critical Survey, from the Greeks to the Present,* expanded edition (Ithaca, N.Y.: Cornell University Press, 1993), 498.

15. Bert O. States, *Great Reckonings in Little Rooms: On the Phenomenology of Theater* (Berkeley: University of California Press, 1987), 48–49.

16. Schmitt, *Actors and Onlookers,* 44.

17. Ibid., 57.

18. I am indebted to Brian Schwartz for pointing out to me the implications of Lorenz's ideas. For an insightful discussion of Gleick's position vis-à-vis Stoppard's central theme of the romantic versus the rational, see Edwards, "Science in *Hapgood* and *Arcadia,*" 178.

19. Vees-Gulani, "Hidden Order," 411.

20. Stoppard, *Arcadia,* 5.

21. N. S. Lockyer, "*Arcadia* and the Physics" (Penn Reading Project instructor seminar, University of Pennsylvania, 25 June 1995).

22. Kramer and Kramer, "Stoppard's *Arcadia,*" 5

23. Stoppard, *Arcadia,* 5.

24. Ibid., 73.

25. Ibid., 15.

26. Ibid., 38.

27. Melbourne, "Plotting the Apple of Knowledge," 557.

28. Stoppard, *Arcadia,* 5.

29. Christopher Innes, "Arcadian Thermodynamics and Chaos Theory" (unpublished paper, Theatres of Science: Crossovers and Confluences, University of Glamorgan, Wales, 9 September 2004).

30. Vees-Gulani, "Hidden Order," 416.

31. Carl Djerassi, "Contemporary 'Science-in-Theatre': A Rare Genre," *Interdisciplinary Science Reviews* 27, no. 3 (Autumn 2002), 194.

32. Stephenson, *Experiment with an Air-Pump,* 29.

33. Melbourne, " 'Plotting the Apple of Knowledge,' " 560, 569.

34. Stephen Abell, "Scratcher of Sores," review of Ben Okri's novel *In Arcadia, Times Literary Supplement* (27 September 2002).

35. Stoppard, *Arcadia,* 93, 78.

36. Ibid., 79.

37. Stuart Curran, notes on the romanticism in *Arcadia* (Penn Reading Project).

38. Lord Byron, "Darkness," in *Works,* 17 vols. (London: John Murray, 1832–33); see also http://eir.library.utoronto.ca/rpo/display/poem3.

39. Jeremy Treglown, "Those Who Can, Teach Also," *Times Literary Supplement* (10 October 1997), 20.

40. Ibid.

41. Ibid.
42. Paul Taylor, review of *The Elephant Vanishes*, Barbican, London, *Independent* (3 July 2003).
43. Theatre de Complicite, *Mnemonic* (London: Methuen, 2001), 3.
44. Ibid. The program notes to the production detail the scientific references, which include John Berger, *A Seventh Man*, plus original material written for *Mnemonic*; Rita Carter, *Mapping the Mind*; and Francis A. Yates, *The Art of Memory*.
45. Theatre de Complicite, *Mnemonic*, 3.
46. Ibid., 3 (ellipsis in the original).
47. Ibid., 3–4.
48. Ibid., 4.
49. Ibid. For further reading on this aspect of memory, see Shepherd-Barr and Shepherd, "Madeleines and Neuromodernism," 39–60.
50. Peter Brook, *The Empty Space* (New York: Atheneum, 1968), 134, 136; emphasis added.
51. Theatre de Complicite, *Mnemonic*, 34.
52. Ibid., 54–55.
53. Ibid., 75.
54. Ibid., 76.
55. Ibid., 75.
56. Brook, *Empty Space*, 140.
57. Ibid., 136.
58. Boyd and Crouch quoted in Paul Taylor, "The Stage We're Going Through," *Independent Review* (10 June 2004), 8.

Chapter Seven: Doctors' Dilemmas: Medicine
under the Scalpel

1. See, for example, the work of D. Heyward Brock, author of "The Doctor as Dramatic Hero," *Perspectives in Biology and Medicine* 34 (Winter 1991), 279–95, and editor of a special issue on "The Doctor and the Drama" in *Literature and Medicine* 12 (Spring 1993), 1.
2. David Jaymes, "Parasitology in Molière: Satire of Doctors and Praise of Paramedics," *Literature and Medicine* 12, no. 1 (Spring 1993), 1–18.
3. McDermott, "Living Newspaper as a Dramatic Form," 89. According to McDermott, *Medicine Show* was not published; it can be consulted among the Living Newspaper Research Materials of the FTP Records in the National Archives.
4. Joanna Howard, foreword to Nell Dunn, *Cancer Tales* (Charlbury, England: Amber Lane Press, 2002).
5. Anna Deavere Smith, performance of an untitled work commissioned for the Yale Tercentennial Celebrations, Yale Medical School (October 2000).
6. David H. Flood and Rhonda L. Soricelli, "The Seventh Chair: An Audience Encounter," *Literature and Medicine* 12, no. 1 (Spring 1993), 43.
7. James M. Welsh, "Strong Medicine at the Movies: A Review," *Literature and Medicine* 12, no. 1 (Spring 1993), 111–20.

8. Henrik Ibsen, *An Enemy of the People*, trans. William Archer (New York: Charles Scribner, 1913), 32.

9. For a thoroughgoing analysis of the play, and especially the character of Stockmann, see Thomas F. Van Laan, "Generic Complexity in Ibsen's *An Enemy of the People*," *Comparative Drama* 20, no. 2 (Summer 1986), 95–114. Stockmann's "problematic qualities" are analyzed on pp. 102–4.

10. Ibsen, *Enemy of the People*, 45.

11. For a more in-depth analysis of the scientific elements of this play, see Asbjørn Aarseth, "Ibsen and Darwin: A Reading of *The Wild Duck*," *Modern Drama* 48, no. 1 (Spring 2005), 1–10.

12. Henrik Ibsen, *A Doll's House*, trans. William Archer, in *The Collected Works of Henrik Ibsen*, vol. 7 (New York: Scribner's, 1906), 81.

13. Jan Lazardzig, "Inszenierung wissenschaftlicher Tatsachen in der Syphilisaufklärung: 'Die Schiffbrüchigen' im Deutschen Theater zu Berlin (1913)," *Der Hautarzt* 53, no. 4 (2002), 268–76.

14. *Three Plays by Brieux*, trans. Mrs. Bernard Shaw, St. John Hankin, and John Pollock, preface by Bernard Shaw (London: A. C. Fifield, 1911), ix.

15. George Bernard Shaw, *The Doctor's Dilemma*, in *Representative Modern Plays: British*, ed. Robert Warnock (Chicago: Scott, Foresman, 1953), xx.

16. Shaw, *Doctor's Dilemma*, p. 191.

17. Ibid. In quoting the play I have retained Shaw's idiosyncratic spelling of contractions.

18. Ibid., 223.

19. Ibid., 192.

20. Sally Peters, "Shaw's Life: A Feminist in Spite of Himself," in *The Cambridge Companion to George Bernard Shaw*, ed. Christopher Innes (Cambridge: Cambridge University Press, 1998), 10–11.

21. Ibid, 11.

22. Ibid.

23. The history of microbiology is well documented in Bruno Latour's study *The Pasteurization of France*, trans. Alan Sheridan and John Law (Cambridge, Mass.: Harvard University Press, 1993).

24. Shaw, *Doctor's Dilemma*, 191.

25. Ibid., 198–99.

26. Ibid., 195.

27. Ibid., 194–95.

28. Ibid., 238–39.

29. Nancy Franklin, "Wit and Wisdom," review of *Wit*, *New Yorker* (18 January 1999), 86.

30. Ibid.

31. Margaret Edson, *Wit* (London: Faber and Faber, 1999), 13–15.

32. Blansfield, "Atom and Eve," 6.

33. Franklin, "Wit and Wisdom," 87.

34. Edson, *Wit*, 81–84.

35. For discussion of this strategy, see Lustig and Shepherd-Barr, "Science as Theater."

36. Franklin, "Wit and Wisdom," 84.
37. Ibid., 87.
38. Edson, *Wit*, 80.
39. Ibid., 50.
40. Ibid.
41. Ibid.
42. Ibid., 49.
43. Oliver Sacks, "The Case of Anna H.," *New Yorker* (7 October 2002), 63–73. Brook and Estienne's play was published by Methuen (2002); Sacks's book that inspired it is published by Peter Smith (1992). *At First Sight*, the film version of Sacks's "To See and Not See," was directed by Irwin Winkler (1999). "The Man Who Learnt to See," a BBC2 documentary (2002), follows the progress of a patient like Molly Sweeney.
44. Christopher Murray, ed., *Brian Friel: Essays, Diaries, Interviews 1964–1999* (London: Faber and Faber, 1999), 157.
45. Ibid., 159.
46. Ibid., 162.
47. Ibid., 155–56. Friel's diary entries during the writing of this play also indicate that he was suffering from his own problems with vision, consulting an ophthalmologist and undergoing an operation on his right eye for cataracts. The recovery proved difficult: "The right eye is bullying me. I think I'm sorry I had the operation," he notes in the midst of completing *Molly Sweeney* (163).
48. Carole-Anne Upton, "Visions of the Sightless in Friel's *Molly Sweeney* and Synge's *The Well of the Saints*," *Modern Drama* 40, no. 3 (Fall 1997), 348.
49. Murray, *Brian Friel*, 163.
50. Ibid., 158.
51. Ibid., 157.
52. Ibid., 156.
53. Brian Friel, *Molly Sweeney* (New York: Plume, 1994), 19, 25.
54. Ibid., 6, 8.
55. Ibid., 44.
56. Ibid., 53.
57. Ibid., 14.
58. Ibid., 47.
59. Ibid.
60. Ibid., 59.
61. Ibid.
62. Ibid., 65.
63. Murray, *Brian Friel*, 158.
64. Brook, *Empty Space*, 221.
65. Ibid., 221–22.
66. There are odd metatheatrical moments in both of these earlier plays, such as Shaw's reference to himself as the characters discuss his ideas in the middle of *The Doctor's Dilemma* and Ibsen's clever implication of the audience as part of the crowd of townspeople when Doctor Stockmann addresses the mob in act 4 of *An Enemy of the People*. However, neither of these moments fully breaks the

fourth wall; they do not involve any actor explicitly acknowledging or addressing the audience as in the later plays discussed.

67. See, for example, Jerome Groopman, "Dying Words," *New Yorker* (28 October 2002), 62–70.

68. Estelle Manette Raban, "*Men in White* and *Yellow Jack* as Mirrors of the Medical Profession," *Literature and Medicine* 12, no. 1 (Spring 1993), 19–41.

CHAPTER EIGHT: "JUST A FICTION": STAGING HISTORY AND TRUTH

1. Wertenbaker, *After Darwin*, 37.
2. Marko Juvan, "On Literature as a Cultural Memory" (proceedings of conference on cultural memory at http://clcwebjournal.lib.purdue.edu/clcweb99–1/juvan 99.htm).
3. Lewis Wolpert, *Good Evening, Galileo* (unpublished play). I am indebted to the author for sharing his manuscript with me.
4. Eric Bentley, introduction to *Galileo*, 12.
5. Ibid., 11.
6. Ibid., 13.
7. Ibid.
8. Kipphardt, *In the Matter of J. Robert Oppenheimer*, 5; quoting Hegel, *Aesthetik* (Berlin, 1955), pt. III, chap. 3, p. 897.
9. Jonathan Culler, *Literary Theory: A Very Short Introduction* (Oxford: Oxford University Press, 1997), 92.
10. Ibid.
11. Bentley, introduction to *Galileo*, 30.
12. Marie-Laure Ryan, "Postmodernism and the Doctrine of Panfictionality," *Narrative* 5 (May 1997), 167–68; emphasis added.
13. Ibid., 166.
14. Ibid.
15. Gerald Holton, letter to the editor, *New York Times* (9 February 2002).
16. Kipphardt, *In the Matter of J. Robert Oppenheimer*, 5. The play's subtitle is *A Play Freely Adapted on the Basis of the Documents*.
17. Ibid.
18. Miller, *Speech Acts in Literature*, 18.
19. Stephen J. Bottoms, "The Efficacy/Effeminacy Braid: Unpicking the Performance Studies/Theatre Studies Dichotomy," *Theatre Topics* 13, no. 2 (September 2003), 180–81.
20. Robert Marc Friedman, "Reflections of a Historian of Science" (the Niels Bohr Archive's History of Science Seminar, Copenhagen, Denmark, 19 November 1999). Proceedings on the Web at http://www.nbi.dk/NBA/files/sem/copfried .html.
21. Jonathan Mandell, "In Depicting History, Just How Far Can the Facts Be Bent?" *New York Times* (3 March 2002), 7.
22. Ibid.
23. Stephen Greenblatt, "Culture," in *Critical Terms for Literary Study*, ed. Frank Lentricchia and Thomas McLaughlin (Chicago: University of Chicago Press, 1995), 230.

24. Ibid., 230–31.
25. Mandell, "In Depicting History," 7.
26. Ibid.
27. Ryan, "Postmodernism and the Doctrine of Panfictionality," 165–66.
28. Ibid., 166.
29. Robert Marc Friedman, "Dangers of Dramatizing Science," *Physics World* (November 2002), 17.
30. Ibid.
31. Ibid.
32. Ibid.
33. Anders Barany, " 'Remembering Miss Meitner,' a Play by Robert Marc Friedman," *Interdisciplinary Science Reviews* 27, no. 3 (2002), 246.
34. Carl Djerassi, keynote presentation, Theatres of Science conference, University of Glamorgan, Wales (9 September 2004).
35. Djerassi, "Contemporary 'Science-in-Theatre,' " 193.
36. Jacket copy review quotations, *Oxygen* (Weinheim, Federal Republic of Germany: Wiley-VCH, 2001).
37. Eva-Sabine Zehelein, "Carl Djerassi's Science-in-Theater: From PUS and PEST to PUSH," unpublished conference presentation, Theatres of Science conference, University of Glamorgan, Wales (8 September 2004).
38. Suzanne Lynch, "A Winning Formula for the Stage?" *Irish Times* (23 November 2001).
39. Ibid. See also Mike Vanden Heuvel, "Playing Science," *RCEI* 50 (2005), 201–12.
40. Ibid.
41. Djerassi, "Contemporary 'Science-in-Theatre,' " 200.
42. Kenneth A. Ribet, review of *Partition* by Ira Hauptmann, *Notices of the American Mathematical Society* 50, no. 11 (December 2003), 1408.
43. Culler, *Literary Theory*, 103.
44. Favorini, "Representation and Reality," 38.
45. Ibid., 38–39.
46. Ibid., 39.
47. P. J. Kavanagh, "Bywords," *Times Literary Supplement* (5 April 2002), 18.

CONCLUSION: ALTERNATING CURRENTS: NEW TRENDS IN SCIENCE AND THEATER

1. Luca Ronconi in an interview with Maria Grazia Gregori (trans. Bruna Tortorella), Sigma Tau Foundation.
2. Ibid.
3. Jean-François Peyret and Alain Prochiantz, *Les Variations Darwin*, program notes (November–December 2004), Théâtre National de Chaillot, trans. Lisbeth Shepherd.
4. I have examined the interaction between form and content in recent science plays in greater depth in the following articles: Lustig and Shephard-Barr, "Science as Theater," 550–55; "*Copenhagen* and Beyond," 171–82; and "From *Copenhagen* to Infinity and Beyond," 193–99.
5. Jean-François Peyret, interview with the author, Cambridge, England (March 2004), trans. Lisbeth Shepherd.

6. Ibid.

7. Heather Neill, "Simon McBurney, Magic Man," *Independent* (23 May 2004).

8. Paul Taylor, review of *The Elephant Vanishes*, Barbican, London, *Independent* (3 July 2003).

9. Janelle Reinelt, review of *Mnemonic*, *Theatre Journal* 52, no. 4 (December 2000), 578–79.

10. See Kirsten Shepherd-Barr, "Acting Out the Search for Infinity," *Physics World* (July 2003), 38–39.

11. Ronconi in an interview with the author, Teatro Piccolo, Milan (May 2003), trans. Pino Donghi.

12. Ibid.

13. Schmitt, *Actors and Onlookers*, 127.

14. Ibid., 116.

15. Ibid.

16. See especially Gillian Beer, "Translation or Transformation? The Relations of Literature and Science," in *Open Fields: Science in Cultural Encounter* (Oxford: Oxford University Press, 1996), 173–95.

17. Complicite, program for *Mnemonic*, 27.

18. Ibid.

19. Agnes de Cayeux, *Les Variations Darwin*, program notes (November–December 2004), Théâtre National de Chaillot, trans. Lisbeth Shepherd.

20. Schmitt, *Actors and Onlookers*, 126.

21. *La Génisse en gésine*, un film de Jacquie Bablet, kindly given to me by Jean-François Peyret and Alain Prochiantz.

22. Peyret and Prochiantz, *Les Variations Darwin*, program notes.

23. Ibid.

24. Ibid.

25. Laura Spinney, "A Stage of Evolution," *Nature* 432 (25 November 2004), 445.

26. Jean-Didier Vincent, interview with author, Hamden, Connecticut (April 2004).

27. Jean-François Peyret, private communication with the author. *Le Cas de Sophie K.* had its premiere in Avignon in July 2005 and will be performed in Paris in April–May 2006.

28. Ibid.

29. Heather Neill, "Simon McBurney, Magic Man," *Independent* (23 May 2004).

30. Alastair Macaulay, "Diamond, What!" review of *The Elephant Vanishes*, *Times Literary Supplement* (1 August 2003).

31. Ibid.

32. Vincent interview.

33. Ibid.

34. Pointedly, his company is called Théâtre Feuilleton (feuilleton means "series").

35. Vincent interview.

36. Neill, "Simon McBurney." Peter Brook expresses the same dissatisfaction with the English term for "audience": "In the French language amongst the different terms for those who watch, for public, for spectator, one word stands out, is different in quality from the rest. *Assistance*—I watch a play: *j'assiste à une pièce*. To assist—the word is simple: it is the key." Brook, *Empty Space*, 139.

37. Shepherd-Barr, *Ibsen and Early Modernist Theatre*, passim.

38. Schmitt, *Actors and Onlookers*, 118.

39. John Brockway Schmor, "Devising New Theater for College Programs," *Theater Topics* 14, no. 1 (March 2004), 259.

40. Michael Vanden Heuvel, "Complementary Spaces: Realism, Performance and a New Dialogics of Theater," *Theater Journal* 44, no. 1 (March 1992), 58.

41. George, "Quantum Theater," 171.

42. Ibid., p. 172.

43. David Barnett, "Reading and Performing Uncertainty: Michael Frayn's *Copenhagen* and the Postdramatic Theatre," *Theatre Research International* 30, no. 2 (Summer 2005), 140.

44. Ibid.

45. Hans-Thies Lehmann, *Postdramatisches Theater* (Frankfurtam Main: Verlag der Autoren, 1999), 330; quoted in Barnett, "Reading and Performing Uncertainty."

46. Christopher Balme, editorial, *Theater Research International* 29, no. 1 (March 2004), 1.

47. Barnett, "Reading and Performing Uncertainty"; emphasis added. It should be noted that Barnett considers *Copenhagen* to be an example of postdramatic theater.

48. Thomas Postlewait, "Theater History and Historiography: A Disciplinary Mandate," *Theater Survey* 45, no. 2 (November 2004), 184.

49. Vanden Heuvel, "Complementary Spaces," 48.

50. Knight, "Scientific Lectures," 217.

51. Taylor, "The Stage We're Going Through," 9. Churchill's contribution is minimal, consisting of a few lines of dialogue from some of her previous work. I would suggest that her short play *A Number* (2002) represents a much more interesting and full engagement with scientific ideas (in this case, cloning and the nature versus nurture debate about identity formation).

52. Lythgoe quoted in Taylor, "The Stage We're Going Through," 9.

53. Ibid.

54. Ibid.

55. It should be noted that many of these pieces dealing with science and medicine in performance have been funded by the Wellcome Trust through its Sciart schemes, which stipulate that applicants avoid traditional proscenium theater.

56. Program notes to *Yerma's Eggs*, by Anna Furse, produced by Athletes of the Heart at Riverside Studios, London, May–June 2004.

57. Beer, *Open Fields*, v.

Bibliography

•••••••••••••••••

Selected Plays

Auburn, David. *Proof.* New York: Faber and Faber, 2001.

Barrow, John. *Infinities.* Milan: Edizioni Piccolo teatro di Milano, 2002.

Brecht, Bertolt. *Galileo.* Trans. Charles Laughton. Ed. Eric Bentley. New York: Grove Press, 1966.

Brenton, Howard. *The Genius.* In *Plays: 2.* London: Methuen, 1989.

Brook, Peter, and Marie-Hélène Estienne. *The Man Who.* London: Methuen, 2002.

Churchill, Caryl. *A Number.* London: Nick Hern Books, 2002.

Davis, Hallie Flanagan. *E = mc²: A Living Newspaper about the Atomic Age.* New York: Samuel French, 1948.

Djerassi, Carl, and Roald Hoffman. *Oxygen.* Weinheim, Federal Republic of Germany: Wiley-VCH, 2001.

Dürrenmatt, Friedrich. *The Physicists.* Trans. James Kirkup. New York: Grove Press, 1964.

Edson, Margaret. *Wit.* New York: Faber and Faber, 1999.

Frayn, Michael. *Copenhagen.* London: Methuen, 2003.

Friel, Brian. *Molly Sweeney.* New York: Plume, 1994.

Ibsen, Henrik. *An Enemy of the People.* Trans. William Archer. New York: Charles Scribner, 1913.

Johnson, Terry. *Insignificance.* In *Plays: 1.* London: Methuen, 1997.

Jonson, Ben. *The Alchemist.* Ed. Simon Trussler. London: Nick Hern Books, 1996.

Kipphardt, Heinar. *In the Matter of J. Robert Oppenheimer.* Trans. Ruth Speirs. New York: Hill and Wang, 1968.

Lawrence, Jerome, and Robert E. Lee. *Inherit the Wind.* New York: Bantam Books, 1960.

MacColl, Ewan. *Uranium 235.* Manchester: Manchester University Press, 1986.

Marlowe, Christopher. *Doctor Faustus.* Ed. David Bevington and Eric Rasmussen. Manchester: Manchester University Press, 1993.

Shaw, George Bernard. *The Doctor's Dilemma.* In *Representative Modern Plays: British,* ed. Robert Warnock. Chicago: Scott, Foresman, 1953.

Stephenson, Shelagh. *An Experiment with an Air-Pump.* London: Methuen, 1998.

Stoppard, Tom. *Arcadia.* London: Faber and Faber, 1993.

———. *Galileo: A Screenplay. Arete* 11 (Spring/Summer 2003).

———. *Hapgood.* London: Faber and Faber, 1988; revised 1994.

Theatre de Complicite. *Mnemonic.* London: Methuen, 2001.

Wertenbaker, Timberlake. *After Darwin.* London: Faber and Faber, 1998.

Wilder, Thornton. *The Skin of our Teeth.* In *Three Plays.* New York: Avon Books, 1976.

Wolpert, Lewis. "Good Evening, Galileo." Unpublished playlet courtesy of the author.

Secondary Works: Books and Articles

Aarseth, Asbjørn. "Ibsen and Darwin: A Reading of *The Wild Duck*." *Modern Drama* 48, no. 1 (Spring 2005), 1–10.

Antor, H. "The Arts, the Sciences and the Making of Meaning: Tom Stoppard's *Arcadia* as a Post-structuralist Play." *Anglia* 116, no. 3 (1998), 326–54.

Arends, Bergit, and Davina Thackara, eds. *Experiment: Conversations in Art and Science*. London: Wellcome Trust, 2003.

Argyros, Alexander. *A Blessed Rage for Order: Deconstruction, Evolution and Chaos*. Ann Arbor: University of Michigan Press, 1991.

Balme, Christopher. Editorial. *Theatre Research International* 29, no. 1 (March 2004), 1–3.

Barba, Eugenio. *The Dilated Body; Followed by the Gospel According to Oxyrhincus*. Rome: Zeami Libri, 1985.

Barrow, John D. "Infinity Upstaged." *New Scientist* 27, no. 2414 (September 2003), 34–37.

Beer, Gillian. *Open Fields: Science in Cultural Encounter*. Oxford: Oxford University Press, 1996.

Bentley, Eric. "The Science Fiction of Bertolt Brecht." In Bertolt Brecht, *Life of Galileo*, 9–42. English version by Charles Laughton. New York: Grove Press, 1966.

Bentley, Joanne. *Hallie Flanagan: A Life in the American Theatre*. New York: Knopf, 1988.

Bernstein, Jeremy. *Hitler's Uranium Club: The Secret Recordings at Farm Hall*. 2nd ed. New York: Copernicus Books, 2001.

Bigsby, C.W.E. *A Critical Introduction to Twentieth-Century American Drama*. Vol. 1, *1900–1940*. Cambridge: Cambridge University Press, 1989.

Blansfield, Karen C. "Atom and Eve: The Mating of Science and Humanism." *South Atlantic Review* 68, no. 4 (Fall 2003), 1–16.

Bottoms, Stephen J. "The Efficacy/Effeminacy Braid: Unpicking the Performance Studies/Theatre Studies Dichotomy." *Theatre Topics* 13, no. 2 (September 2003), 173–87.

Brecht, Bertolt. *Brecht on Theatre: The Development of an Aesthetic*. Ed. and trans. John Willett. New York: Hill and Wang, 1964.

Brook, Peter. *The Empty Space*. New York: Atheneum, 1968.

———. *Threads of Time: A Memoir*. London: Methuen, 1998.

Burge, Jim. "In Search of Science on the Big Screen." *Interdisciplinary Science Reviews* 27, no. 3 (Autumn 2002), 165–68.

Campos, Liliane. "La Métaphore scientifique dans le théâtre britannique contemporain." M.Phil. thesis. Université Paris 4—Sorbonne, September 2004.

Carlson, Marvin. *The Haunted Stage: The Theatre as Memory Machine*. Ann Arbor: University of Michigan Press, 2001.

———. *Performance: A Critical Introduction*. London: Routledge, 1993.

———. *Theories of the Theatre: A Historical and Critical Survey, from the Greeks to the Present*. Expanded edition. Ithaca, N.Y.: Cornell University Press, 1993.

Carpenter, Charles A. *Dramatists and the Bomb: American and British Playwrights Confront the Nuclear Age, 1945–1964*. Westport, Conn.: Greenwood Press, 1999.

Cassidy, David C. "Heisenberg, Uncertainty and the Quantum Revolution." *Scientific American* (May 1992), 106–12.

Cattermole, Howard, ed. *Interdisciplinary Science Reviews*: 27, no. 3 (Autumn 2002), 161–247. Special issue on science and theater.

Clurman, Harold. *The Collected Works of Harold Clurman: Six Decades of Commentary on Theatre, Dance, Music, Film, Arts, and Letters*. Ed. Marjorie Roggia and Glenn Young. New York: Applause Books, 1994.

Cornwell, John. *Hitler's Scientists: Science, War, and the Devil's Pact*. New York: Viking, 2003.

Crockett, Roger A. *Understanding Friedrich Dürrenmatt*. Columbia: University of South Carolina Press, 1998.

Culler, Jonathan. *Literary Theory: A Very Short Introduction*. Oxford: Oxford University Press, 1997.

Dasenbrock, Reed. *Truth and Consquences: Intentions, Conventions, and the New Thematics*. Pittsburgh: Pennsylvania State University Press, 2000.

De Beer, Gavin, ed. *Charles Darwin and Thomas Henry Huxley: Autobiographies*. Oxford: Oxford University Press, 1983.

Demastes, William W. *Theatre of Chaos: Beyond Absurdism, into Orderly Disorder*. Cambridge: Cambridge University Press, 1998.

Djerassi, Carl. "Contemporary 'Science-in-Theatre': A Rare Genre." *Interdisciplinary Science Reviews* 27, no. 3 (Autumn 2002), 193–201.

Duffy, Susan, and Bernard K. Duffy. "Theatrical Responses to Technology during the Depression: Three Federal Theatre Project Plays." *Theatre History Studies* 6 (1986), 142–64.

Esslin, Martin. "'Dead! And Never Called Me Mother!': The Missing Dimension in American Drama." *Studies in the Literary Imagination* (Fall 1988), 23–33.

Fahy, Thomas, and Kimball King, eds. *Peering behind the Curtain: Disability, Illness and the Extraordinary Body in Contemporary Theater*. New York: Routledge, 2002.

Farrington, Benjamin. *What Darwin Really Said*. New York: Schocken, 1966.

Favorini, Attilio. "Representation and Reality: The Case of Documentary Theatre." *Theatre Survey* 35, no. 2 (November 1994), 31–43.

Fickert, Kurt J. "The Curtain Speech in Dürrenmatt's *The Physicists*." *Modern Drama* 13 (1970), 40–46.

———. *To Heaven and Back: The New Morality in the Plays of Friedrich Dürrenmatt*. Lexington: University Press of Kentucky, 1972.

Fine, Arthur. *The Shaky Game: Einstein, Realism and the Quantum Theory*. Chicago: University of Chicago Press, 1986.

Fischer-Seidel, Therese. "Chaos Theory, Landscape Gardening, and Tom Stoppard's Dramatology of Coincidence in *Arcadia*." In *Emerging Structures in Interdisciplinary Perspective*, ed. Rudi Keller and Karl Menges. Tübingen: Francke, 1997.

Fleming, John. *Stoppard's Theatre: Finding Order amid Chaos*. Austin: University of Texas Press, 2001.

Foreman, Richard. *Unbalancing Acts: Foundations for a Theater*. Ed. Ken Jordan. New York: Theatre Communications Group, 1993.

Frayn, Michael. "'Copenhagen' Revisited." *New York Review of Books* (28 March 2002), 22–24.

Freeman, Sara. "Adaptation after Darwin: Timberlake Wertenbaker's Evolving Texts." *Modern Drama* 45, no. 2 (2002), 646–62.

Friedman, Robert Marc. "Dangers of Dramatizing Science." *Physics World* (November 2002), 16–17.

———. " 'Remembering Miss Meitner': An attempt to Forge History into Drama." *Interdisciplinary Science Reviews* 27, no. 3 (Autumn 2002), 202–10.

Furey, Marguerite. "The Responsibility of the Scientist as a Theme in Modern German Drama." M.A. thesis, Duke University, 1967.

Gargano, Cara. "Complex Theatre: Science and Myth in Three Contemporary Performances." *New Theatre Quarterly* 14, no. 2 (May 1998), 151–58.

Glanz, James. "Details of Nazi's A-Bomb Program Surface." *New York Times* (7 January 2002).

———. "Frayn Takes Stock of Bohr Revelations." *New York Times* (9 February 2002).

———. "New Twist on Physicist's Role in Nazi Bomb." *New York Times* (7 February 2002), A1, A12.

———. "Secret Trove May Resolve 'Copenhagen.'" *New York Times* (20 October 2001).

Goodall, Jane R. *Performance and Evolution in the Age of Darwin: Out of the Natural Order.* London: Routledge, 2002.

Goorney, Howard, and Ewan MacColl. *Agit-Prop to Theatre Workshop: Political Playscripts 1930–1950.* Manchester: Manchester University Press, 1986.

Gould, Stephen Jay. *Ever since Darwin: Reflections in Natural History.* New York: Norton, 1992.

———. *Full House: The Spread of Excellence from Plato to Darwin.* New York: Three Rivers Press, 1997.

———. *The Structure of Evolutionary Theory.* Cambridge, Mass.: Belknap Press, 2002.

Greenblatt, Stephen. "Culture." In *Critical Terms for Literary Study*, ed. Frank Lentricchia and Thomas McLaughlin, 225–32. 2nd ed. Chicago: University of Chicago Press, 1995.

Greene, Brian. *The Elegant Universe: Superstrings, Hidden Dimensions, and the Quest for the Ultimate Theory.* London: Vintage, 2000.

Gussow, Mel. *Theatre on the Edge: New Visions, New Voices.* New York: Applause, 1998.

Harding, Sandra. *Is Science Multicultural? Post-colonialism, Feminism, and Epistemologies.* Bloomington: Indiana University Press, 1998.

———. *The Science Question in Feminism.* Ithaca, N.Y.: Cornell University Press, 1986.

———. *Whose Science? Whose Knowledge? Thinking from Women's Lives.* Ithaca, N.Y.: Cornell University Press, 1991.

Hawkins, Harriett. *Strange Attractors: Literature, Culture, and Chaos Theory.* New York: Prentice Hall/Harvester Wheatsheaf, 1995.

Hayles, N. Katherine, ed. *Chaos and Order: Complex Dynamics in Literature and Science.* Chicago: University of Chicago Press, 1991.

———. *Chaos Bound: Orderly Disorder in Contemporary Literature and Science.* Ithaca, N.Y.: Cornell University Press, 1990.

———. *The Cosmic Web: Scientific Field Models and Literary Strategies in the Twentieth Century.* Ithaca, N.Y.: Cornell University Press, 1984.

Haynes, Roslynn D. *From Faust to Strangelove: Representations of the Scientist in Western Literature*. Baltimore: Johns Hopkins University Press, 1994.

Hays, Michael. "Dramatic Literature as History: Some Suggestions about Theory and Method." *Bucknell Review* 23, no. 2 (1977), 131–41.

Heisenberg, Martin. "With the Bomb for the Bomb." Trans. Grethe Shepherd. Letter to Editor. *Die Zeit* (14 March 2002).

Hellen, Nicholas. "Hitler's Bomb Chief Betrayed Nuclear Secret." *Sunday Times* [London] (6 January 2002).

Hentschel, Klaus. "What History of Science Can Learn from Michael Frayn's 'Copenhagen.'" *Interdisciplinary Science Reviews* 27, no. 3 (Autumn 2002), 211–16.

Hustvedt, Asti. "Science Fictions: Villiers de l'Isle Adam's 'L'Ève Future' and Late Nineteenth-Century Medical Constructions of Femininity." Ph.D. diss., New York University, 1996.

Huxley, Aldous. *Literature and Science*. New York: Harper and Row, 1963.

Hye, Allen E. *The Moral Dilemma of the Scientist in Modern Drama*. Lewiston, N.Y.: Edwin Mellen Press, 1996.

Innes, Christopher, ed. *The Cambridge Companion to George Bernard Shaw*. Cambridge: Cambridge University Press, 1998.

James, Clive. "Count Zero Splits the Infinite." In *Critical Essays on Tom Stoppard*, ed. Anthony Jenkins, 27–34. London: G. K. Hall, 1990.

Jenkins, Anthony. "Moles and Molecules: Tom Stoppard's *Hapgood*." In *Critical Essays on Tom Stoppard*, ed. Anthony Jenkins, 164–74. London: G. K. Hall, 1990.

Jernigan, D. "Tom Stoppard and 'Post-Modern Science': Normalizing Radical Epistemologies in *Hapgood* and *Arcadia*." *Comparative Drama* 37 (2003), 3–35.

Jordanova, L. J., ed. *Languages of Nature: Critical Essays on Science and Literature*. Foreword by Raymond Williams. London: Free Association, 1986.

Karlsch, Rainer. *Hitlers Bombe*. Munich: Deutsche Verlags-Anstalt, 2005.

Kavanagh, P. J. "Bywords." *Times Literary Supplement* (5 April 2002), 18.

Kelly, Katherine E. *The Cambridge Companion to Tom Stoppard*. Cambridge: Cambridge University Press, 2001.

———. *Tom Stoppard and the Craft of Comedy*. Ann Arbor: University of Michigan Press, 1991.

Kramer, Prapassaree, and Jeffrey Kramer. "Stoppard's *Arcadia*: Research, Time, Loss." *Modern Drama* 40, no. 1 (Spring 1997), 1–10.

Krance, Charles. "Odd Fizzles: Beckett and the Heavenly Sciences." *Bucknell Review: Science and Literature* 27, no. 2 (1983), 96–107.

Kuberski, Philip. *Chaosmos: Literature, Science, and Theory*. Albany: State University of New York Press, 1994.

Kuhn, Thomas S. *The Structure of Scientific Revolutions*. 2nd ed. Chicago: University of Chicago Press, 1970.

Latour, Bruno. *Science in Action*. Cambridge, Mass.: Harvard University Press, 1988.

Latour, Bruno, and Steve Woolgar. *Laboratory Life: The Construction of Scientific Facts*. Princeton, N.J.: Princeton University Press, 1979.

Levine, George, ed. *Realism and Representation: Essays on the Problem of Realism in Relation to Science, Literature and Culture*. Madison: University of Wisconsin Press, 1993.

Levine, George, ed., with the assistance of Alan Rauch. *One Culture: Essays in Science and Literature*. Madison: University of Wisconsin Press, 1987.

Lightman, Alan. "Art That Transfigures Science." *New York Times* (15 March 2003).

Lustig, Harry, and Kirsten Shepherd-Barr. "Science as Theater." *American Scientist* 90 (November–December 2002), 550–55.

Lyon, James K. *Bertolt Brecht in America*. Princeton, N.J.: Princeton University Press, 1980.

MacColl, Ewan. *Journeyman: An Autobiography by Ewan MacColl*. London: Sidgwick and Jackson, 1990.

Malekin, Peter, and Ralph Yarrow. *Consciousness, Literature and Theatre: Theory and Beyond*. London: Macmillan, 1997.

Mandell, Jonathan. "In Depicting History, Just How Far Can the Facts Be Bent?" *New York Times* (3 March 2002), 7–9.

Martin, J. R. *Reading Science: Critical and Functional Perspectives on Discourses of Science*. London: Routledge, 1998.

Martin, Matthew. "Stephen Poliakoff's Drama for the Post-scientific Age." *Theatre Journal* 45, no. 2 (May 1993), 197–211.

May, Leila. "Monkeys, Microcephalous Idiots, and the Barbarous Races of Mankind: Darwin's Dangerous Victorianism." *Victorian Newsletter* (Fall 2002), 20–27.

McDermott, Douglas. "The Living Newspaper as a Dramatic Form." *Modern Drama* 8, no. 1 (1965), 82–94.

Melbourne, Lucy. "'Plotting the Apple of Knowledge': Tom Stoppard's *Arcadia* as Iterated Theatrical Algorithm." *Modern Drama* 41, no. 4 (Winter 1998), 557–72.

Miller, J. Hillis. *Speech Acts in Literature*. Stanford, Calif.: Stanford University Press, 2001.

Morton, Peter. *The Vital Science: Biology and the Literary Imagination, 1860–1900*. London: Allen and Unwin, 1984.

Murphy, Patrick D., ed. *Staging the Impossible: The Fantastic Mode in Modern Drama*. Westport, Conn.: Greenwood Press, 1992.

Murray, Christopher, ed. *Brian Friel: Essays, Diaries, Interviews 1964–1999*. London: Faber and Faber, 1999.

Nellhaus, Tobin. "Science, History, Theatre: Theorizing in Two Alternatives to Positivism." *Theatre Journal* 45, no. 4 (December 1993), 505–27.

Neubauer, John. "Models for the History of Science and of Literature." *Bucknell Review: Science and Literature* 27, no. 2 (1983), 17–37.

Orthofer, M. A. "The Scientist on the Stage: A Survey." *Interdisciplinary Science Reviews* 27, no. 3 (Autumn 2002), 173–83.

Overbye, Dennis. "Lab Coat Chic: The Arts Embrace Science." *New York Times* (28 January 2003).

Pais, Abraham. *Niels Bohr's Times, in Physics, Philosophy, and Polity*. Oxford: Clarendon Press, 1991.

Paulsell, Patricia R. "Brecht's Treatment of the Scientific Method in His Leben des Galilei." *German Studies Review* 11, no. 2 (May 1988), 267–84.

Paulson, William. *Literary Culture in a World Transformed: A Future for the Humanities*. Ithaca, N.Y.: Cornell University Press, 2001.

Paradis, James, and Thomas Postlewait, eds. *Victorian Science and Victorian Values: Literary Perspectives*. New York: New York Academy of Sciences, 1981.

Peppard, Murray B. *Friedrich Dürrenmatt*. New York: Twayne, 1969.

Peterfreund, Stuart, ed. *Literature and Science: Theory and Practice*. Boston: Northeastern University Press, 1990.

Peters, Sally. "Shaw's Life: A Feminist in Spite of Himself." In *The Cambridge Companion to George Bernard Shaw*, ed. Christopher Innes, 3–24. Cambridge: Cambridge University Press, 1998.

Plotnitsky, Arkady. *The Knowable and the Unknowable: Modern Science, Nonclassical Thought, and the Two Cultures*. Ann Arbor: University of Michigan Press, 2002.

Powers, Thomas. "A Letter from Copenhagen." *New York Review of Books* (14 August 2003).

Reinelt, Janelle. "Performance and the Documentary Impulse." Unpublished paper for ASTR seminar on documentary theater, 2001. Part of a larger project entitled *Public Performances: Race and Nation in the Theatre of Our Time*, forthcoming from University of Michigan Press.

Renner, Pamela. "Science and Sensibility." *American Theatre* 16, no. 4 (1999), 34–36.

Rennert, Hal H. "The Threat of the Invisible: The Portrait of the Physicist in Modern German Drama." In *To Hold a Mirror to Nature: Dramatic Images and Reflections*, ed. Karelisa Hartigan, 91–104. Washington, D.C.: University Press of America, 1982.

Reno, Robert P. "Science and Prophets in Dürrenmatt's *The Physicists*." *Renascence: Essays on Value in Literature* 37, no. 2 (Winter 1985), 70–79.

Roach, Joseph. *The Player's Passion: Studies in the Science of Acting*. Newark: University of Delaware Press, 1985.

Roberts, David. "Brecht and the Idea of a Scientific Theatre." *Brecht Yearbook* (1984), 13:41–60.

Rocamora, Carol. "Scientific Dramaturgy." *Nation* (5 June 2000), 49–51.

Rodgers, Peter. "Novel Approaches to Science." *Physics World* (November 2002), 38–39.

———. "Physics Meets Art and Literature." *Physics World* (November 2002), 29.

———. "When Science Takes to the Stage." *Physics World* (November 2002), 32–33.

Roe, Ian F. "Dürrenmatt's Die Physiker: Die drei Leben des Galilei?" *Forum for Modern Language Studies* 27, no. 3 (July 1991), 255–67.

Ross, Andrew, ed. *Science Wars*. Durham, N.C.: Duke University Press, 1996.

Ruddick, Nick. "The Search for a Quantum Ethics: Michael Frayn's *Copenhagen* and Other Recent British Science Plays." *Journal of the Fantastic in the Arts* 11, no. 4 (2001), 415–31.

Ryan, Marie-Laure. "Postmodernism and the Doctrine of Panfictionality." *Narrative* 5 (May 1997), 165–87.

Sacks, Oliver. "To See and Not See." In *An Anthropologist on Mars*, 108–52. New York: Knopf, 1995.

Schleifer, Ronald. "Analogy and Example: Heisenberg, Negation, and the Language of Quantum Physics." *Criticism* 33, no. 3 (Summer 1991), 285–307.

———. *Modernism and Time: The Logic of Abundance in Literature, Science, and Culture, 1880–1930*. Cambridge: Cambridge University Press, 2000.

Schmitt, Natalie Crohn. *Actors and Onlookers: Theater and Twentieth-Century Scientific Views of Nature*. Evanston, Ill.: Northwestern University Press, 1990.

———. "Theorizing about Performance: Why Now?" *New Theatre Quarterly* 6, no. 23 (August 1990), 231–34.

Serres, Michel. *The Natural Contract*. Trans. William Paulson and Elizabeth MacArthur. Ann Arbor: University of Michigan Press, 1995.

———. *The Troubadour of Knowledge*. Trans. Sheila Faria Gloser and William Paulson. Ann Arbor: University of Michigan Press, 1997.

Serres, Michel, with Bruno Latour. *Conversations on Science, Culture, and Time*. Trans. Roxanne Lapidus. Ann Arbor: University of Michigan Press, 1995.

Shaffer, Elinor S., ed. *The Third Culture: Literature and Science*. Berlin: W. de Gruyter, 1998.

Shepherd-Barr, Kirsten. "*Copenhagen* and Beyond: The 'Rich and Mentally Nourishing' Interplay of Science and Theatre." *Gramma: Journal of Theory and Criticism* 10 (2002), 171–82.

———. "*Copenhagen*, Infinity, and Beyond: Science Meets Literature on the Stage." *Interdisciplinary Science Reviews* 28, no. 3 (Fall 2003), 193–99.

———. *Ibsen and Early Modernist Theatre, 1890–1900*. Westport, Conn.: Greenwood Press, 1997.

———. "Mise-en-Scent: The Théâtre d'Art's *Song of Songs* and the Use of Smell as a Theatrical Device." *Theatre Research International* 24, no. 2 (Summer 1999), 152–59.

Shepherd-Barr, Kirsten, and Gordon M. Shepherd. "Madeleines and Neuromodernism: Reassessing Mechanisms of Autobiographical Memory in Proust." *A/B: Autobiography Studies* 13, no. 1 (Spring 1998), 39–60.

Shlain, Leonard. *Art and Physics: Parallel Visions in Space, Time and Light*. New York: Perennial, 1993.

Slade, Joseph W., and Judith Yaross Lee, eds. *Beyond the Two Cultures: Essays on Science, Technology, and Literature*. Ames: Iowa State University Press, 1990.

Smith, Murray. "Darwin and the Directors: Film, Emotion and the Face in the Age of Evolution." *Times Literary Supplement* (7 February 2003), 13–15.

Snow, C. P. *The Physicists*. Introduction by William Cooper. Boston: Little, Brown, 1981.

———. *The Two Cultures and a Second Look*. Cambridge: Cambridge University Press, 1984.

Solomon, Alisa. *Re-dressing the Canon: Essays on Theater and Gender*. New York: Routledge, 1997.

Soupel, Serge, and Roger A. Hambridge. *Literature and Science and Medicine: Papers Read at the Clark Library Summer Seminar 1981*. Los Angeles: William Andrews Clark Memorial Library, 1982.

States, Bert O. *Great Reckonings in Little Rooms: On the Phenomenology of Theater*. Berkeley: University of California Press, 1987.

Stewart, Victoria. "A Theatre of Uncertainties: Science and History in Michael Frayn's 'Copenhagen.' " *New Theatre Quarterly* 15 (November 1999), 301–7.

Styan, J. L. *Modern Drama in Theory and Practice*. 3 vols. Cambridge: Cambridge University Press, 1983.

Vanden Heuvel, Michael. "Complementary Spaces: Realism, Performance and a New Dialogics of Theatre." *Theatre Journal* 44, no. 1 (March 1992), 47–58.

———. *Performing Drama/Dramatizing Performance: Alternative Theater and the Dramatic Text.* Ann Arbor: University of Michigan Press, 1991.

———. "Playing Science." *Revista Canaria de Estudios Ingleses* 50 (April 2005), 201–12.

Vanhoutte, J. "Cancer and the Common Woman in Margaret Edson's *Wit*." *Comparative Drama* 36, nos. 3–4 (2002), 391–410.

Vees-Gulani, Susanne. "Hidden Order in the 'Stoppard Set': Chaos Theory in the Content and Structure of Tom Stoppard's *Arcadia*." *Modern Drama* 42, no. 3 (Fall 1999), 411–26.

Weber, Bruce. "Science Finding a Home Onstage." *New York Times* (2 June 2000), E1, E7.

Wilcox, Dean. "What Does Chaos Theory Have to Do with Art?" *Modern Drama* 39 (1996), 698–711.

Wilson, Stephen. *Information Arts: Intersections of Art, Science, and Technology.* Cambridge, Mass.: MIT Press, 2002.

SELECTED THEATER REVIEWS

Barany, Anders. "'Remembering Miss Meitner,' a Play by Robert Marc Friedman." *Interdisciplinary Science Reviews* 27, no. 3 (Autumn 2002), 245–46.

Billington, Michael. "Limitless Possibilities." *Guardian* (27 November 1999).

———. "A Number." *Guardian* (27 September 2002), 23.

Blakemore, Michael. "From Physics to Metaphysics and the Bomb." *New York Times* (9 April 2000), 7, 20.

Brantley, Ben. "Into the Loop of a Daisy Chain of Memories." *New York Times* (29 March 2001), E1, E5.

Coyaud, Sylvie. Review of John Barrow's *Infinities. Interdisciplinary Science Reviews* 27, no. 3 (Autumn 2002), 246–47.

Franklin, Nancy. "Wit and Wisdom." Review of *Wit*. *New Yorker* (18 January 1999), 86–87.

Gardner, Lyn. "To Bee or Not to Bee, That Is the Question" [about *Humble Boy*, by Charlotte Jones]. *Guardian* (18 July 2001).

Gee, Maggie. "Copenhagen." *Times Literary Supplement* (12 June 1998), 19.

Gill, Michael. Review of *After Darwin. Socialist Standard* (October 1998), http://www.worldsocialism.org/spgb/oct98/theatre.html.

Hoffman, Roald, and Sylvie Coyaud. "Infinite Ideas." *Nature* 416 (11 April 2002), 585.

Kozinn, Allan. "A Heretical Astonomer Rethinking His Revolution." *New York Times* (3 October 2002).

Lahr, John. "Bombs and Qualms." *New Yorker* (24 April and 1 May 2000), 219–20.

McBurney, Simon. "You Must Remember This." *Guardian* (1 January 2003).

Nightingale, Benedict. Review of *After Darwin* and other productions in London, *New York Times* (9 August 1998).

Panek, Richard. "Discovering Drama, Even Song, in Dry Old Science." *New York Times* (29 September 2002), Arts and Leisure, 1, 30.

Schechner, Richard. *Performance Studies: An Introduction.* New York: Routledge, 2002.

Shepherd-Barr, Kirsten. "Hilbert's Hotel, Other Paradoxes, Come to Life in New 'Math Play.'" *SIAM NEWS* 36, no. 7 (September 2003), 1, 7.

Spinney, Laura. "A Stage of Evolution." *Nature* 432 (25 November 2004), 445.

Tollini, Frederick P. "Blessed Uncertainty." *America* (23 September 2000), 24–25.

Index

●●●●●●●●●●